W9-BNN-298

TO MY WIFE

Bibliographical Note

This Dover edition, first published in 1995, is an unabridged republication of the 9th printing, 1961, of the 1946 second, revised edition of the work originally published by Oxford University Press, New York, in 1941.

Library of Congress Cataloging-in-Publication Data

Tarski, Alfred.
[O logice matematycznej i metodzie dedukcyjnej. English]
Introduction to logic and to the methodology of deductive sciences / Alfred Tarski ; translated by Olaf Helmer.
p. cm.
Originally published: 2nd ed., rev. New York : Oxford University Press, 1946 (1961 printing).
Includes bibliographical references (p.) and index.
ISBN-13: 978-0-486-28462-0
ISBN-10: 0-486-28462-X
1. Mathematics—Philosophy. 2. Arithmetic—Foundations. I. Title.
QA9.T28 1995
511.3—dc20 94-48412
CIP

Manufactured in the United States by LSC Communications
28462X14 2020
www.doverpublications.com

INTRODUCTION TO LOGIC

AND TO THE

METHODOLOGY OF DEDUCTIVE SCIENCES

ALFRED TARSKI

TRANSLATED BY OLAF HELMER

DOVER PUBLICATIONS, IN[illegible]

New York

TABLE OF CONTENTS

IV. On the Theory of Classes

V. On the Theory of Relations

VI. On the Deductive Method

Second Part

APPLICATIONS OF LOGIC AND METHODOLOGY IN CONSTRUCTING MATHEMATICAL THEORIES

VII. Construction of a Mathematical Theory: Laws of Order for Numbers

VIII. Construction of a Mathematical Theory: Laws of Addition and Subtraction

IX. Methodological Considerations on the Constructed Theory

X. Extension of the Constructed Theory. Foundations of Arithmetic of Real Numbers

PREFACE

The present book is a partially modified and extended edition of my book *On Mathematical Logic and Deductive Method*, which appeared first in 1936 in Polish and then in 1937 in an exact German translation (under the title: *Einführung in die mathematische Logik und in die Methodologie der Mathematik*). In its original form it was intended as a popular scientific book; its aim was to present to the educated layman—in a manner which would combine scientific exactitude with the greatest possible intelligibility—a clear idea of that powerful trend of contemporary thought which is concentrated about modern logic. This trend arose originally from the somewhat limited task of stabilizing the foundations of mathematics. In its present phase, however, it has much wider aims. For it seeks to create a unified conceptual apparatus which would supply a common basis for the whole of human knowledge. Furthermore, it tends to perfect and sharpen the deductive method, which in some sciences is regarded as the sole permitted means of establishing truths, and indeed in every domain of intellectual activity is at least an indispensable auxiliary tool for deriving conclusions from accepted assumptions.

The response accorded to the Polish and German editions, and especially some suggestions made by reviewers, gave rise to the idea of making the new edition not merely a popular scientific book, but also a textbook upon which an elementary college course in logic and the methodology of deductive sciences could be based. The experiment seemed the more desirable in view of a certain lack of suitable elementary textbooks in this domain.

In order to carry out the experiment, it was necessary to make several changes in the book.

Some very fundamental questions and notions were entirely passed over or merely touched upon in the previous editions, either because of their more technical character, or in order to avoid points of a controversial nature. As examples may be cited such topics as the difference between the usage of certain logical notions in systematic developments of logic and in the language of everyday life, the general method of verifying the laws of the sentential calculus, the necessity of a sharp distinction between words and their names, the concepts of the universal class and the null class, the fundamental notions of the calculus of relations, and finally the conception of methodology as a general science of sciences. In the present edition all these topics are discussed (although not all in an equally thorough manner), since it seemed to me that to avoid them would constitute an essential gap in any textbook of modern logic. Consequently, the chapters of the first, general part of the book have been more or less extended; in particular, Chapter II, which is devoted to the sentential calculus, contains much new material. I have also added many new exercises to these chapters, and have increased the number of historical indications.

While in previous editions the use of special symbols was reduced to a minimum, I considered it necessary in the present edition to familiarize the reader with the elements of logical symbolism. Nevertheless, the use of this symbolism in practice remains very restricted, and is limited mostly to exercises.

In previous editions the principal domain from which examples were drawn for illustrating general and abstract considerations was high-school mathematics; for it was, and still is, my opinion that elementary mathematics, and especially algebra, because of the simplicity of its concepts and the uniformity of its methods of inference, is peculiarly appropriate for exemplifying various fundamental phenomena of a logical and methodological nature. Nevertheless, in the present edition, particularly in the newly added passages, I draw examples more frequently from other domains, especially from everyday life.

Independent of these additions, I have rewritten certain sections whose mastery by students had been found somewhat difficult.

The essential features of the book remain unchanged. The preface to the original edition, the major part of which is reprinted in the next few pages, will give the reader an idea of the general character of the book. Perhaps, however, it is desirable to point out explicitly at this place what he should not expect to find in it.

First, the book contains no systematic and strictly deductive presentation of logic; such a presentation would obviously not lie within the framework of an elementary textbook. It was originally my intention to include, in the present edition, an additional chapter entitled *Logic as a Deductive Science*, which—as an illustration of the general methodological remarks contained in Chapter VI—would outline a systematic development of some elementary parts of logic. For a number of reasons this intention could not be realized; but I hope that several new exercises on this subject included in Chapter VI will to some extent compensate for the omission.

Secondly, apart from two rather short passages, the book gives no information about the traditional Aristotelian logic, and contains no material drawn from it. But I believe that the space here devoted to traditional logic corresponds well enough to the small role to which this logic has been reduced in modern science; and I also believe that this opinion will be shared by most contemporary logicians.

And, finally, the book is not concerned with any problems belonging to the so-called logic and methodology of empirical sciences. I must say that I am inclined to doubt whether any special "logic of empirical sciences," as opposed to logic in general or the "logic of deductive sciences," exists at all (at least so long as the word "logic" is used as in the present book—that is to say, as the name of a discipline which analyzes the meaning of the concepts common to all the sciences, and establishes the general laws governing the concepts). But this is rather a terminological, than a factual, problem. At any rate the methodology of empirical sciences constitutes an important domain of scientific research. The knowledge of logic is of course valuable in the study of this methodology, as it is in the case of any other discipline. It must be admitted, however, that logical concepts and methods have not, up to the present, found any specific or fertile applications in this domain. And it is at least possible that this situation is not merely a consequence of the present stage of methodological researches. It arises, perhaps, from the circumstance that, for the purpose of an adequate methodological treatment, an empirical science may have to be considered, not merely as a scientific theory —that is, as a system of asserted statements arranged according to certain rules—, but rather as a complex consisting partly of such statements and partly of human activities. It should be added that, in striking opposition to the high development of the empirical sciences themselves, the methodology of these sciences can hardly boast of comparably definite achievements—despite the great efforts that have been made. Even the preliminary task of clarifying the concepts involved in this domain has not yet been carried out in a satisfactory way. Consequently, a course in the methodology of empirical sciences must have a quite different character from one in logic and must be largely confined to evaluations and criticisms of tentative gropings and unsuccessful efforts. For these and other reasons, I see little rational justification for combining the discussion of logic and the methodology of empirical sciences in the same college course.

A few remarks concerning the arrangement of the book and its use as a college text.

The book is divided into two parts The first gives a general introduction to logic and the methodology of deductive sciences; the second shows, by means of a concrete example, the sort of applications which logic and methodology find in the construction of mathematical theories, and thus affords an opportunity to assimilate and deepen the knowledge acquired in the first part. Each chapter is followed by appropriate exercises. Brief historical indications are contained in footnotes.

Passages, and even whole sections, which are set off by asterisks "*" both at the beginning and at the end, contain more difficult material, or presuppose familiarity with other passages containing such material; they can be omitted without jeopardizing the intelligibility of subsequent parts of the book. This also applies to the exercises whose numbers are preceded by asterisks.

I feel that the book contains sufficient material for a full-year course. Its arrangement, however, makes it feasible to use it in half-year courses as well. If used as a text in half-year logic courses in a department of philosophy, I suggest the thorough study of its first part, including the more difficult portions, with the entire omission of the second part. If the book is used in a half-year course in a mathematics department—for instance, in the foundations of mathematics—, I suggest the study of both parts of the book, with the omission of the more difficult passages.

In any case, I should like to emphasize the importance of working out the exercises carefully and thoroughly; for they not only facilitate the assimilation of the concepts and principles discussed, but also touch upon many problems for the discussion of which the text provided no opportunity.

I shall be very happy if this book contributes to the wider diffusion of logical knowledge. The course of historical events has assembled in this country the most eminent representatives of contemporary logic, and has thus created here especially favorable conditions for the development of logical thought. These favorable conditions can, of course, be easily overbalanced by other and more powerful factors. It is obvious that the future of logic, as well as of all theoretical science, depends essentially upon normalizing the political and social relations of mankind, and thus upon a factor which is beyond the control of professional scholars. I have no illusions that the development of logical thought, in particular, will have a very essential effect upon the process of the normalization of human relationships; but I do believe that the wider diffusion of the knowledge of logic may contribute positively to the acceleration of this process. For, on the one hand, by making the meaning of concepts precise and uniform in its own field and by stressing the necessity of such a precision and uniformization in any other domain, logic leads to the possibility of better understanding between those who have the will to do so. And, on the other hand, by perfecting and sharpening the tools of thought, it makes men more critical—and thus makes less likely their being misled by all the pseudo-reasonings to which they are in various parts of the world incessantly exposed today.

I gratefully acknowledge my indebtedness to Dr. O. Helmer, who performed the translation of the German edition into English.

I want to express my warmest gratitude to Dr. A. Hofstadter, Mr. L. K. Krader, Professor E. Nagel, Professor W. V. Quine, Mr. M. G. White and especially Dr. J. C. C. McKinsey and Dr. P. P. Wiener, who were unsparing in their advice and assistance while I was preparing the English edition. I also owe many thanks to Mr. K. J. Arrow for his help in reading proofs.

Alfred Tarski

Harvard University September 1940

The present book is a photographic reprint of the first English edition, and no large-scale changes could be introduced in it. Misprints, however, have been corrected, and a number of improvements in detail have been made. I wish to thank readers and reviewers for their helpful suggestions, and I am especially indebted to Miss Louise H. Chin for her assistance in preparing the present edition for publication.

A. T.

University of California,
Berkeley, August 1945

FROM THE PREFACE TO THE ORIGINAL EDITION

In the opinion of many laymen mathematics is today already a dead science: after having reached an unusually high degree of development, it has become petrified in rigid perfection. This is an entirely erroneous view of the situation; there are but few domains of scientific research which are passing through a phase of such intensive development at present as mathematics. Moreover, this development is extraordinarily manifold: mathematics is expanding its domain in all possible directions, it is growing in height, in width, and in depth. It is growing in height, since, on the soil of its old theories which look back upon hundreds if not thousands of years of development, new problems appear again and again, and ever more perfect results are being achieved. It is growing in width, since its methods permeate other branches of sciences, while its domain of investigation embraces increasingly more comprehensive ranges of phenomena and ever new theories are being included in the large circle of mathematical disciplines. And finally it is growing in depth, since its foundations become more and more firmly established, its methods perfected, and its principles stabilized.

It has been my intention in this book to give those readers who are interested in contemporary mathematics, without being actively concerned with it, at least a very general idea of that third line of mathematical development, i.e. its growth in depth. My aim has been to acquaint the reader with the most important concepts of a discipline which is known as mathematical logic, and which has been created for the purpose of a firmer and more profound establishment of the foundations of mathematics; this discipline, in spite of its brief existence of barely a century, has already attained a high degree of perfection and plays today a role in the totality of our knowledge that far transcends its originally intended boundaries. It has been my intention to show that the concepts of logic permeate the whole of mathematics, that they comprehend all specifically mathematical concepts as special cases, and that logical laws are constantly applied—be it consciously or unconsciously—in mathematical reasonings. Finally, I have tried to present the most important principles in the construction of mathematical theories—principles which form the subject matter of still another discipline, the methodology of mathematics—and to show how one sets about using those principles in practice.

It has not been easy to carry this whole plan through within the framework of a relatively small book without presupposing on the part of the reader any specialized mathematical knowledge or any specific training in reasonings of an abstract character. Throughout the book a combination of the greatest possible intelligibility with the necessary conciseness had to be attempted, with a constant care for avoiding errors or cruder inexactitudes from the scientific standpoint. A language had to be used which deviates as little as possible from the language of everyday life. The employment of a special logical symbolism had to be given up, although this symbolism is an invaluable tool which permits us to combine conciseness with precision, removes to a large degree the possibility of ambiguities and misunderstandings, and is thereby of essential service in all subtler considerations. The idea of a systematic treatment had to be abandoned from the beginning. Of the abundance of questions which present themselves only a few could be discussed in detail, others could only be touched upon superficially, while still others had to be passed over entirely, with the consciousness that the selection of topics discussed would inevitably exhibit a more or less arbitrary character. In those cases in which contemporary science has as yet not taken any definite stand and offers a number of possible and equally correct solutions, it was out of the question to present objectively all known views. A decision in favor of a definite point of view had to be made. When making such a decision I have taken care, not primarily to have it conform to my personal inclinations, but rather to choose a method of solution which would be as simple as possible and which would lend itself to a popular mode of presentation.

I do not have the illusion that I have entirely succeeded in overcoming these and other difficulties.

FIRST PART

•

ELEMENTS OF LOGIC

DEDUCTIVE METHOD

· I ·

ON THE USE OF VARIABLES

1. Constants and variables

Every scientific theory is a system of sentences which are accepted as true and which may be called LAWS or ASSERTED STATEMENTS or, for short, simply STATEMENTS. In mathematics, these statements follow one another in a definite order according to certain principles which will be discussed in detail in Chapter VI, and they are, as a rule, accompanied by considerations intended to establish their validity. Considerations of this kind are referred to as PROOFS, and the statements established by them are called THEOREMS.

Among the terms and symbols occurring in mathematical theorems and proofs we distinguish CONSTANTS and VARIABLES. In arithmetic, for instance, we encounter such constants as "*number*", "*zero*" ("0"), "*one*" ("1"), "*sum*" ("+"), and many others.[1] Each of these terms has a well-determined meaning which remains unchanged throughout the course of the considerations.

As variables we employ, as a rule, single letters, e.g. in arithmetic the small letters of the English alphabet: "*a*", "*b*", "*c*",

[1] By "arithmetic" we shall here understand that part of mathematics which is concerned with the investigation of the general properties of numbers, relations between numbers and operations on numbers. In place of the word "arithmetic" the term "algebra" is frequently used, particularly in high-school mathematics. We have given preference to the term "arithmetic" because, in higher mathematics, the term "algebra" is reserved for the much more special theory of algebraic equations. (In recent years the term "algebra" has obtained a wider meaning, which is, however, still different from that of "arithmetic".)—The term "number" will here always be used with that meaning which is normally attached to the term "real number" in mathematics; that is to say, it will cover integers and fractions, rational and irrational, positive and negative numbers, but not imaginary or complex numbers.

. . ., "x", "y", "z". As opposed to the constants, the variables do not possess any meaning by themselves. Thus, the question:

does zero have such and such a property?

e.g.:

is zero an integer?

can be answered in the affirmative or in the negative; the answer may be true or false, but at any rate it is meaningful. A question concerning x, on the other hand, for example the question:

is x an integer?

cannot be answered meaningfully.

In some textbooks of elementary mathematics, particularly the less recent ones, one does occasionally come across formulations which convey the impression that it is possible to attribute an independent meaning to variables. Thus it is said that the symbols "x", "y", · · · also denote certain numbers or quantities, not "constant numbers" however (which are denoted by constants like "0", "1", · · ·), but the so-called "variable numbers" or rather "variable quantities". Statements of this kind have their source in a gross misunderstanding. The "variable number" x could not possibly have any specified property, for instance, it could be neither positive nor negative nor equal to zero; or rather, the properties of such a number would change from case to case, that is to say, the number would sometimes be positive, sometimes negative, and sometimes equal to zero. But entities of such a kind we do not find in our world at all; their existence would contradict the fundamental laws of thought. The classification of the symbols into constants and variables, therefore, does not have any analogue in the form of a similar classification of the numbers.

2. Expressions containing variables—sentential and designatory functions

In view of the fact that variables do not have a meaning by themselves, such phrases as:

x is an integer

are not sentences, although they have the grammatical form of sentences; they do not express a definite assertion and can be neither confirmed nor refuted. From the expression:

x is an integer

we only obtain a sentence when we replace "x" in it by a constant denoting a definite number; thus, for instance, if "x" is replaced by the symbol "1", the result is a true sentence, whereas a false sentence arises on replacing "x" by "$\frac{1}{2}$". An expression of this kind, which contains variables and, on replacement of these variables by constants, becomes a sentence, is called a SENTENTIAL FUNCTION. But mathematicians, by the way, are not very fond of this expression, because they use the term "function" with a different meaning. More often the word "CONDITION" is employed in this sense; and sentential functions and sentences which are composed entirely of mathematical symbols (and not of words of everyday language), such as:

$$x + y = 5,$$

are usually referred to by mathematicians as FORMULAS. In place of "sentential function" we shall sometimes simply say "sentence" —but only in cases where there is no danger of any misunderstanding.

The role of the variables in a sentential function has sometimes been compared very adequately with that of the blanks left in a questionnaire; just as the questionnaire acquires a definite content only after the blanks have been filled in, a sentential function becomes a sentence only after constants have been inserted in place of the variables. The result of the replacement of the variables in a sentential function by constants—equal constants taking the place of equal variables—may lead to a true sentence; in that case, the things denoted by those constants are said to SATISFY the given sentential function. For example, the numbers 1, 2 and $2\frac{1}{2}$ satisfy the sentential function:

$$x < 3,$$

but the numbers 3, 4 and $4\frac{1}{2}$ do not.

Besides the sentential functions there are some further expressions containing variables that merit our attention, namely, the so-called DESIGNATORY OR DESCRIPTIVE FUNCTIONS. They are expressions which, on replacement of the variables by constants, turn into designations ("descriptions") of things. For example, the expression:

$$2x + 1$$

is a designatory function, because we obtain the designation of a certain number (e.g., the number 5), if in it we replace the variable "x" by an arbitrary numerical constant, that is, by a constant denoting a number (e.g., "2").

Among the designatory functions occurring in arithmetic, we have, in particular, all the so-called algebraic expressions which are composed of variables, numerical constants and symbols of the four fundamental arithmetical operations, such as:

$$x - y, \qquad \frac{x+1}{y+2}, \qquad 2\cdot(x + y - z).$$

Algebraic equations, on the other hand, that is to say, formulas consisting of two algebraic expressions connected by the symbol "$=$", are sentential functions. As far as equations are concerned, a special terminology has become customary in mathematics; thus, the variables occurring in an equation are referred to as the unknowns, and the numbers satisfying the equation are called the roots of the equation. E.g., in the equation:

$$x^2 + 6 = 5x$$

the variable "x" is the unknown, while the numbers 2 and 3 are roots of the equation.

Of the variables "x", "y", $\cdots$ employed in arithmetic it is said that they STAND FOR DESIGNATIONS OF NUMBERS or that numbers are VALUES of these variables. Thereby approximately the following is meant: a sentential function containing the symbols "x", "y", $\cdots$ becomes a sentence, if these symbols are replaced by such constants as designate numbers (and not by expressions designating operations on numbers, relations between numbers or even things outside the field of arithmetic like geometrical configurations, animals, plants, etc.). Likewise, the variables occurring in geometry stand for designations of points and geometrical figures. The designatory functions which we meet in arithmetic can also be said to stand for designations of numbers. Sometimes it is simply said that the symbols "x", "y", $\cdots$ themselves, as well as the designatory functions made up out of them, denote numbers or are designations of numbers, but this is then a merely abbreviative terminology.

3. Formation of sentences by means of variables—universal and existential sentences

Apart from the replacement of variables by constants there is still another way in which sentences can be obtained from sentential functions. Let us consider the formula:

$$x + y = y + x.$$

It is a sentential function containing the two variables "x" and "y" that is satisfied by any arbitrary pair of numbers; if we put any numerical constants in place of "x" and "y", we always obtain a true formula. We express this fact briefly in the following manner:

for any numbers x and y, $x + y = y + x.$

The expression just obtained is already a genuine sentence and, moreover, a true sentence; we recognize in it one of the fundamental laws of arithmetic, the so-called commutative law of addition. The most important theorems of mathematics are formulated similarly, namely, all so-called UNIVERSAL SENTENCES, or SENTENCES OF A UNIVERSAL CHARACTER, which assert that arbitrary things of a certain category (e.g., in the case of arithmetic, arbitrary numbers) have such and such a property. It has to be noticed that in the formulation of universal sentences the phrase "*for any things* (or *numbers*) x, y, $\cdots$ " is often omitted and has to be inserted mentally; thus, for instance, the commutative law of addition may simply be given in the following form:

$$x + y = y + x.$$

This has become a well accepted usage, to which we shall generally adhere in the course of our further considerations.

Let us now consider the sentential function:

$$x > y + 1.$$

This formula fails to be satisfied by every pair of numbers; if, for instance, "3" is put in place of "x" and "4" in place of "y", the false sentence:

$$3 > 4 + 1$$

is obtained. Therefore, if one says:

$$\textit{for any numbers } x \textit{ and } y, \quad x > y + 1,$$

one does undoubtedly state a meaningful, though obviously false, sentence. There are, on the other hand, pairs of numbers which satisfy the sentential function under consideration; if, for example, "x" and "y" are replaced by "4" and "2", respectively, the result is the true formula:

$$4 > 2 + 1.$$

This situation is expressed briefly by the following phrase:

$$\textit{for some numbers } x \textit{ and } y, \quad x > y + 1,$$

or, using a more frequently employed form:

$$\textit{there are numbers } x \textit{ and } y \textit{ such that} \quad x > y + 1.$$

The expressions just given are true sentences; they are examples of EXISTENTIAL SENTENCES, OR SENTENCES OF AN EXISTENTIAL CHARACTER, stating the existence of things (e.g., numbers) with a certain property.

With the help of the methods just described we can obtain sentences from any given sentential function; but it depends on the content of the sentential function whether we arrive at a true or a false sentence. The following example may serve as a further illustration. The formula:

$$x = x + 1$$

is satisfied by no number; hence, no matter whether the words "*for any number x*" or "*there is a number x such that*" are prefixed, the resulting sentence will be false.

In contradistinction to sentences of a universal or existential character we may denote sentences not containing any variables, such as:

$$3 + 2 = 2 + 3,$$

as SINGULAR SENTENCES. This classification is not at all exhaustive, since there are many sentences which cannot be counted among any of the three categories mentioned. An example is represented by the following statement:

for any numbers x and y there is a number z such that
$$x = y + z.$$

Sentences of this type are sometimes called CONDITIONALLY EXISTENTIAL SENTENCES (as opposed to the existential sentences considered before, which may also be called ABSOLUTELY EXISTENTIAL SENTENCES); they state the existence of numbers having a certain property, but on condition that certain other numbers exist.

4. Universal and existential quantifiers; free and bound variables

Phrases like:

for any x, y, ⋯

and

there are x, y, ⋯ such that

are called QUANTIFIERS; the former is said to be a UNIVERSAL, the latter an EXISTENTIAL QUANTIFIER. Quantifiers are also known as OPERATORS; there are, however, expressions counted likewise among operators, which are different from quantifiers. In the preceding section we tried to explain the meaning of both quantifiers. In order to emphasize their significance it may be pointed out that, only by the explicit or implicit employment of operators, can an expression containing variables occur as a sentence, that is, as the statement of a well-determined assertion.

Without the help of operators, the usage of variables in the formulation of mathematical theorems would be excluded.

In everyday language it is not customary (though quite possible) to use variables, and quantifiers are also, for this reason, not in use. There are, however, certain words in general usage which exhibit a very close connection with quantifiers, namely, such words as "*every*", "*all*", "*a certain*", "*some*". The connection becomes obvious when we observe that expressions like:

all men are mortal

or

some men are wise

have about the same meaning as the following sentences, formulated with the help of quantifiers:

for any x, if x is a man, then x is mortal

and

there is an x, such that x is both a man and wise,

respectively.

For the sake of brevity, the quantifiers are sometimes replaced by symbolic expressions. We can, for instance, agree to write in place of:

for any things (or *numbers*) $x, y, \cdots$

and

there exist things (or *numbers*) $x, y, \cdots$ *such that*

the following symbolic expressions:

$$\mathsf{A}_{x,y,\cdots} \quad \text{and} \quad \mathsf{E}_{x,y,\cdots}$$

respectively (with the understanding that the sentential functions following the quantifiers are put in parentheses). According to this agreement, the statement which was given at the end of the preceding section as an example of a conditionally existential sentence, for instance, assumes the following form:

$$\text{(I)} \qquad \mathsf{A}_{x,y}[\mathsf{E}_{z}(x = y + z)]$$

A sentential function in which the variables "x", "y", "z", ... occur automatically becomes a sentence as soon as one prefixes to it one or several operators containing all those variables. If, however, some of the variables do not occur in the operators, the expression in question remains a sentential function, without becoming a sentence. For example, the formula:

$$x = y + z$$

changes into a sentence if preceded by one of the phrases:

for any numbers x, y and z;

there are numbers x, y and z such that;

for any numbers x and y, there is a number z such that;

and so on. But if we merely prefix the quantifier:

there is a number z such that or $\underset{z}{\mathsf{E}}$

we do not yet arrive at a sentence; the expression obtained, namely:

(II) $$\underset{z}{\mathsf{E}}(x = y + z)$$

is, however, undoubtedly a sentential function, for it immediately becomes a sentence when we substitute some constants in the place of "x" and "y" and leave "z" unaltered, or else, when we prefix another suitable quantifier, e.g.:

for any numbers x and y or $\underset{x,y}{\mathsf{A}}$

It is seen from this that, among the variables which may occur in a sentential function, two different kinds can be distinguished. The occurrence of variables of the first kind—they will be called FREE OR REAL VARIABLES—is the decisive factor in determining that the expression under consideration is a sentential function and not a sentence; in order to effect the change from a sentential function to a sentence it is necessary to replace these variables by constants or else to put operators in front of the sentential function that contain those free variables. The remaining, so-called BOUND OR APPARENT VARIABLES, however, are not to be changed in such a transformation. In the above sentential function (II), for instance, "x" and "y" are free variables, and the symbol "z" occurs twice as a bound variable; on the other hand, the expression (I) is a sentence, and thus contains bound variables only.

*It depends entirely upon the structure of the sentential function, namely, upon the presence and position of the operators, whether any particular variable occurring in it is free or bound. This may be best seen by means of a concrete example. Let us, for instance, consider the following sentential function:

(III) *for any number x, if* $x = 0$ *or* $y \neq 0$, *then*
there exists a number z such that $x = y \cdot z$.

This function begins with a universal quantifier containing the variable "x", and therefore the latter, which occurs three times in this function, occurs at all these places as a bound variable; at the first place it makes up part of the quantifier, while at the other two places it is, as we say, BOUND BY THE QUANTIFIER. The situation is similar with respect to the variable "z". For, although the initial quantifier of (III) does not contain this variable, we can, nevertheless, recognize a certain sentential function forming a part of (III) which opens with an existential quantifier containing the variable "z"; this is the function:

(IV) *there exists a number z such that* $x = y \cdot z$.

Both places at which the variable "z" occurs in (III) belong to the partial function (IV) just stated. It is for this reason that we say that "z" occurs everywhere in (III) as a bound variable; at the first place it makes up part of the existential quantifier, and at the second place it is bound by that quantifier. As for the variable "y" also occurring in (III), we see that there is no quantifier in (III) containing this variable, and therefore it occurs in (III) twice as a free variable.

The fact that quantifiers bind variables—that is, that they change free into bound variables in the sentential functions which follow them—constitutes a very essential property of quantifiers. Several other expressions are known which have an analogous property; with some of them we shall become acquainted later (in Sections 20 and 22), while some others—such as, for instance, the integral sign—play an important role in higher mathematics. The term "operator" is the general term used to denote all expressions having this property.*

5. The importance of variables in mathematics

As we have seen in Section 3 variables play a leading role in the formulation of mathematical theorems. From what has been said it does not follow, however, that it would be impossible in principle to formulate the latter without the use of variables. But in practice it would scarcely be feasible to do without them, since even comparatively simple sentences would assume a complicated and obscure form. As an illustration let us consider the following theorem of arithmetic:

for any numbers x and y, $x^3 - y^3 = (x - y) \cdot (x^2 + xy + y^2)$.

Without the use of variables, this theorem would look as follows:

> *the difference of the third powers of any two numbers is equal to the product of the difference of these numbers and a sum of three terms, the first of which is the square of the first number, the second the product of the two numbers, and the third the square of the second number.*

An even more essential significance, from the standpoint of the economy of thought, attaches to variables as far as mathematical proofs are concerned. This fact will be readily confirmed by the reader if he attempts to eliminate the variables in any of the proofs which he will meet in the course of our further considerations. And it should be pointed out that these proofs are much simpler than the average considerations to be found in the various fields of higher mathematics; attempts at carrying the latter through without the help of variables would meet with very considerable difficulties. It may be added that it is to the introduction of variables that we are indebted for the development of so fertile a method for the solution of mathematical problems as the method of equations. Without exaggeration it can be said that the invention of variables constitutes a turning point in the history of mathematics; with these symbols man acquired a tool that prepared the way for the tremendous development of the mathematical science and for the solidification of its logical foundations.[2]

Exercises

1. Which among the following expressions are sentential functions, and which are designatory functions:

(a) *x is divisible by* 3,

(b) *the sum of the numbers x and* 2,

(c) $y^2 - z^2$,

(d) $y^2 = z^2$,

(e) $x + 2 < y + 3$,

(f) $(x + 3) - (y + 5)$,

(g) *the mother of x and z*,

(h) *x is the mother of z* ?

2. Give examples of sentential and designatory functions from the field of geometry.

3. The sentential functions which are encountered in arithmetic and which contain only one variable (which may, however, occur at several different places in the given sentential function) can be divided into three categories: (i) functions satisfied by every number; (ii) functions not satisfied by any number; (iii) functions satisfied by some numbers, and not satisfied by others.

[2] Variables were already used in ancient times by Greek mathematicians and logicians,—though only in special circumstances and in rare cases. At the beginning of the 17th century, mainly under the influence of the work of the French mathematician F. Vieta (1540–1603), people began to work systematically with variables and to employ them consistently in mathematical considerations. Only at the end of the 19th century, however, due to the introduction of the notion of a quantifier, was the role of variables in scientific language and especially in the formulation of mathematical theorems fully recognized; this was largely the merit of the outstanding American logician and philosopher Ch. S. Peirce (1839–1914).

To which of these categories do the following sentential functions belong:

(a) $x + 2 = 5 + x$,

(b) $x^2 = 49$,

(c) $(y + 2) \cdot (y - 2) < y^2$,

(d) $y + 24 > 36$,

(e) $z = 0$ *or* $z < 0$ *or* $z > 0$,

(f) $z + 24 > z + 36$?

4. Give examples of universal, absolutely existential and conditionally existential theorems from the fields of arithmetic and geometry.

5. By writing quantifiers containing the variables "x" and "y" in front of the sentential function:

$$x > y$$

it is possible to obtain various sentences from it, for instance:

for any numbers x and y, $x > y$;

for any number x, there exists a number y such that $x > y$;

there is a number y such that, for any number x, $x > y$.

Formulate them all (there are six altogether) and determine which of them are true.

6. Do the same as in Exercise 5 for the following sentential functions:

$$x + y^2 > 1$$

and

x is the father of y

(assuming that the variables "x" and "y" in the latter stand for names of human beings).

7. State a sentence of everyday language that has the same meaning as:

for every x, if x is a dog, then x has a good sense of smell

and that contains no quantifier or variables.

8. Replace the sentence:

some snakes are poisonous

by one which has the same meaning but is formulated with the help of quantifiers and variables.

9. Differentiate, in the following expressions, between the free and bound variables:

(a) *x is divisible by y;*

(b) *for any x,* $x - y = x + (-y)$,

(c) *if* $x < y$, *then there is a number z such that* $x < y$ *and* $y < z$;

(d) *for any number y, if* $y > 0$, *then there is a number z such that* $x = y \cdot z$;

(e) *if* $x = y^2$ *and* $y > 0$, *then, for any number z,* $x > -z^2$;

(f) *if there exists a number y such that* $x > y^2$, *then, for any number z,* $x > -z^2$.

Formulate the above expressions by replacing the quantifiers by the symbols introduced in Section 4.

*10. If, in the sentential function (e) of the preceding exercise, we replace the variable "z" in both places by "y", we obtain an expression in which "y" occurs in some places as a free and in others as a bound variable; in what places and why?

(In view of some difficulties in operating with expressions in which the same variable occurs both bound and free, some logicians prefer to avoid the use of such expressions altogether and not to treat them as sentential functions.)

*11. Try to state quite generally under which conditions a variable occurs at a certain place of a given sentential function as a free or as a bound variable.

12. Which numbers satisfy the sentential function:

there is a number y such that $x = y^2$,

and which satisfy:

there is a number y such that $x \cdot y = 1$?

· II ·

ON THE SENTENTIAL CALCULUS

6. Logical constants; the old logic and the new logic

The constants with which we have to deal in every scientific theory may be divided into two large groups. The first group consists of terms which are specific for a given theory. In the case of arithmetic, for instance, they are terms denoting either individual numbers or whole classes of numbers, relations between numbers, operations on numbers, etc.; the constants which we used in Section 1 as examples belong here among others. On the other hand, there are terms of a much more general character occurring in most of the statements of arithmetic, terms which are met constantly both in considerations of everyday life and in every possible field of science, and which represent an indispensable means for conveying human thoughts and for carrying out inferences in any field whatsoever; such words as "*not*", "*and*", "*or*", "*is*", "*every*", "*some*" and many others belong here. There is a special discipline, namely LOGIC, considered the basis for all the other sciences, whose concern it is to establish the precise meaning of such terms and to lay down the most general laws in which these terms are involved.

Logic developed into an independent science long ago, earlier even than arithmetic and geometry. And yet it has only been recently—after a long period of almost complete stagnation—that this discipline has begun an intensive development, in the course of which it has undergone a complete transformation with the effect of assuming a character similar to that of the mathematical disciplines; in this new form it is known as MATHEMATICAL OR DEDUCTIVE OR SYMBOLIC LOGIC, and sometimes it is also called LOGISTIC. The new logic surpasses the old in many respects,—not only because of the solidity of its foundations and the perfection of the methods employed in its development, but mainly on account of the wealth of concepts and theorems that have been established. Fundamentally, the old traditional logic forms only a fragment of the new, a fragment moreover which, from the point of view of the requirements of other sciences, and of mathematics in particular, is entirely insignificant. Thus, in regard to the aim which we here have, there will in this whole book be but very little opportunity to draw the material for our considerations from traditional logic.[1]

7. Sentential calculus; negation of a sentence, conjunction and disjunction of sentences

Among the terms of a logical character there is a small distinguished group, consisting of such words as "*not*", "*and*", "*or*", "*if* ... , *then* ...". All these words are well-known to us from everyday language, and serve to build up compound sentences from simpler ones. In grammar, they are counted among the so-called sentential conjunctions. If only for this reason, the presence of these terms does not represent a specific property of any particular science. To establish the meaning and usage of these terms is the task of the most elementary and fundamental part of logic, which is called SENTENTIAL CALCULUS, or sometimes PROPOSITIONAL CALCULUS or (less happily) THEORY OF DEDUCTION.[2]

[1] Logic was created by ARISTOTLE, the great Greek thinker of the 4th century B.C. (384–322); his logical writings are collected in the work *Organon*. As the creator of mathematical logic we have to look upon the great German philosopher and mathematician of the 17th century G. W. LEIBNIZ (1646–1716). However, the logical works of LEIBNIZ failed to have a great influence upon the further development of logical investigations; there was even a period in which they sank into oblivion. A continuous development of mathematical logic began only towards the middle of the 19th century, namely at the time when the logical system of the English mathematician G. BOOLE was published (1815–1864; principal work: *An Investigation of the Laws of Thought*, London 1854). So far the new logic has found its most perfect expression in the epochal work of the great contemporary English logicians A. N. WHITEHEAD and B. RUSSELL: *Principia Mathematica* (Cambridge, 1910–1913).

[2] The historically first system of sentential calculus is contained in the work *Begriffsschrift* (Halle 1879) of the German logician G. FREGE (1848–1925) who, without doubt, was the greatest logician of the 19th century. The eminent contemporary Polish logician and historian of logic J. LUKASIEWICZ succeeded in giving sentential calculus a particularly simple and precise form and caused extensive investigations concerning this calculus.

We will now discuss the meaning of the most important terms of sentential calculus.

With the help of the word *"not"* one forms the NEGATION of any sentence; two sentences, of which the first is a negation of the second, are called CONTRADICTORY SENTENCES. In senténtial calculus, the word *"not"* is put in front of the whole sentence, whereas in everyday language it is customary to place it with the verb; or should it be desirable to have it at the beginning of the sentence, it must be replaced by the phrase *"it is not the case that"*. Thus, for example, the negation of the sentence:

1 *is a positive integer*

reads as follows:

1 *is not a positive integer,*

or else:

it is not the case that 1 *is a positive integer.*

Whenever we utter the negation of a sentence, we intend to express the idea that the sentence is false; if the sentence is actually false, its negation is true, while otherwise its negation is false.

The joining of two sentences (or more) by the word *"and"* results in their so-called CONJUNCTION OR LOGICAL PRODUCT; the sentences joined in this manner are called the MEMBERS OF THE CONJUNCTION or the FACTORS OF THE LOGICAL PRODUCT. If, for instance, the sentences:

2 *is a positive integer*

and

$$2 < 3$$

are joined in this way, we obtain the conjunction:

2 *is a positive integer and* $2 < 3$.

The stating of the conjunction of two sentences is tantamount to stating that both sentences of which the conjunction is formed are true. If this is actually the case, then the conjunction is true, but if at least one of its members is false, then the whole conjunction is false.

By joining sentences by means of the word *"or"* one obtains the DISJUNCTION of those sentences, which is also called the LOGICAL SUM; the sentences forming the disjunction are called the MEMBERS OF THE DISJUNCTION or the SUMMANDS OF THE LOGICAL SUM. The word *"or"*, in everyday language, possesses at least two different meanings. Taken in the so-called NON-EXCLUSIVE meaning, the disjunction of two sentences merely expresses that at least one of these sentences is true, without saying anything as to whether or not both sentences may be true; taken in another meaning, the so-called EXCLUSIVE one, the disjunction of two sentences asserts that one of the sentences is true but that the other is false. Suppose we see the following notice put up in a bookstore:

Customers who are teachers or college students are entitled to a special reduction.

Here the word *"or"* is undoubtedly used in the first sense, since it is not intended to refuse the reduction to a teacher who is at the same time a college student. If, on the other hand, a child has asked to be taken on a hike in the morning and to a theater in the afternoon, and we reply:

no, we are going on a hike or we are going to the theater,

then our usage of the word *"or"* is obviously of the second kind since we intend to comply with only one of the two requests. In logic and mathematics, the word *"or"* is always used in the first, non-exclusive meaning; the disjunction of two sentences is considered true if both or at least one of its members are true, and otherwise false. Thus, for instance, it may be asserted:

every number is positive or less than 3,

although it is known that there are numbers which are both positive and less than 3. In order to avoid misunderstandings, it would be expedient, in everyday as well as in scientific language, to use the word *"or"* by itself only in the first meaning, and to replace it by the compound expression "*either* ... *or* ..." whenever the second meaning is intended.

* Even if we confine ourselves to those cases in which the word "*or*" occurs in its first meaning, we find quite noticeable differences between the usages of it in everyday language and in logic. In common language, two sentences are joined by the word "*or*" only when they are in some way connected in form and content. (The same applies, though perhaps to a lesser degree, to the usage of the word "*and*".) The nature of this connection is not quite clear, and a detailed analysis and description of it would meet with considerable difficulties. At any rate, anybody unfamiliar with the language of contemporary logic would presumably be little inclined to consider such a phrase as:

$$2 \cdot 2 = 5 \quad \textit{or} \quad \textit{New York is a large city}$$

as a meaningful expression, and even less so to accept it as a true sentence. Moreover, the usage of the word "*or*" in everyday English is influenced by certain factors of a psychological character. Usually we affirm a disjunction of two sentences only if we believe that one of them is true but wonder which one. If, for example, we look upon a lawn in normal light, it will not enter our mind to say that the lawn is green or blue, since we are able to affirm something simpler and, at the same time, stronger, namely that the lawn is green. Sometimes even, we take the utterance of a disjunction as an admission by the speaker that he does not know which of the members of the disjunction is true. And if we later arrive at the conviction that he knew at the time that one—and, specifically, which—of the members was false, we are inclined to look upon the whole disjunction as a false sentence, even should the other member be undoubtedly true. Let us imagine, for instance, that a friend of ours, upon being asked when he is leaving town, answers that he is going to do so today, tomorrow or the day after. Should we then later ascertain that, at that time, he had already decided to leave the same day, we shall probably get the impression that we were deliberately misled and that he told us a lie.

The creators of contemporary logic, when introducing the word "*or*" into their considerations, desired, perhaps unconsciously, to simplify its meaning and to render the latter clearer and independent of all psychological factors, especially of the presence or absence of knowledge. Consequently, they extended the usage of the word "*or*", and decided to consider the disjunction of any two sentences as a meaningful whole, even should no connection between their contents or forms exist; and they also decided to make the truth of a disjunction—like that of a negation or conjunction—dependent only and exclusively upon the truth of its members. Therefore, a man using the word "*or*" in the meaning of contemporary logic will consider the expression given above:

$$2 \cdot 2 = 5 \quad \textit{or} \quad \textit{New York is a large city}$$

as a meaningful and even a true sentence, since its second part is surely true. Similarly, if we assume that our friend, who was asked about the date of his departure, used the word "*or*" in its strict logical meaning, we shall be compelled to consider his answer as true, independent of our opinion as to his intentions.*

8. Implication or conditional sentence; implication in material meaning

If we combine two sentences by the words "*if* ..., *then* ...", we obtain a compound sentence which is denoted as an IMPLICATION or a CONDITIONAL SENTENCE. The subordinate clause to which the word "*if*" is prefixed is called ANTECEDENT, and the principal clause introduced by the word "*then*" is called CONSEQUENT. By asserting an implication one asserts that it does not occur that the antecedent is true and the consequent is false. An implication is thus true in any one of the following three cases: (i) both antecedent and consequent are true, (ii) the antecedent is false and the consequent is true, (iii) both antecedent and consequent are false; and only in the fourth possible case, when the antecedent is true and the consequent is false, the whole implication is false. It follows that, whoever accepts an implication as true, and at the same time accepts its antecedent as true, cannot but accept its consequent; and whoever accepts an implication as true and rejects its consequent as false, must also reject its antecedent.

* As in the case of disjunction, considerable differences between the usages of implication in logic and everyday language manifest themselves. Again, in ordinary language, we tend to join two sentences by the words "*if* ···, *then* ···" only when there is some connection between their forms and contents. This connection is hard to characterize in a general way, and only sometimes is its nature relatively clear. We often associate with this connection the conviction that the consequent follows necessarily from the antecedent, that is to say, that if we assume the antecedent to be true we are compelled to assume the consequent, too, to be true (and that possibly we can even deduce the consequent from the antecedent on the basis of some general laws which we might not always be able to quote explicitly). Here again, an additional psychological factor manifests itself; usually we formulate and assert an implication only if we have no exact knowledge as to whether or not the antecedent and consequent are true. Otherwise the use of an implication seems unnatural and its sense and truth may raise some doubt.

The following example may serve as an illustration. Let us consider the law of physics:

every metal is malleable,

and let us put it in the form of an implication containing variables:

if x is a metal, then x is malleable.

If we believe in the truth of this universal law, we believe also in the truth of all its particular cases, that is, of all implications obtainable by replacing "x" by names of arbitrary materials such as iron, clay or wood. And, indeed, it turns out that all sentences obtained in this way satisfy the conditions given above for a true implication; it never happens that the antecedent is true while the consequent is false. We notice, further, that in any of these implications there exists a close connection between the antecedent and the consequent, which finds its formal expression in the coincidence of their subjects. We are also convinced that, assuming the antecedent of any of these implications, for instance, "*iron is a metal*", as true, we can deduce from it its consequent "*iron is malleable*", for we can refer to the general law that every metal is malleable.

Nevertheless, some of the sentences discussed just now seem artificial and doubtful from the point of view of common language. No doubt is raised by the universal implication given above, or by any of its particular cases obtained by replacing "x" by the name of a material of which we do not know whether it is a metal or whether it is malleable. But if we replace "x" by "*iron*", we are confronted with a case in which the antecedent and consequent are undoubtedly true; and we shall then prefer to use, instead of an implication, an expression such as:

since iron is a metal, it is malleable.

Similarly, if for "x" we substitute "*clay*", we obtain an implication with a false antecedent and a true consequent, and we shall be inclined to replace it by the expression:

although clay is not a metal, it is malleable.

And finally, the replacement of "x" by "*wood*" results in an implication with a false antecedent and a false consequent; if, in this case, we want to retain the form of an implication, we should have to alter the grammatical form of the verbs:

if wood were a metal, then it would be malleable.

The logicians, with due regard for the needs of scientific languages, adopted the same procedure with respect to the phrase "*if* ..., *then* ..." as they had done in the case of the word "*or*". They decided to simplify and clarify the meaning of this phrase, and to free it from psychological factors. For this purpose they extended the usage of this phrase, considering an implication as a meaningful sentence even if no connection whatsoever exists between its two members, and they made the truth or falsity of an implication dependent exclusively upon the truth or falsity of the antecedent and consequent. To characterize this situation briefly, we say that contemporary logic uses IMPLICATIONS IN MATERIAL MEANING, or simply, MATERIAL IMPLICATIONS; this is opposed to the usage of IMPLICATION IN FORMAL MEANING

OR FORMAL IMPLICATION, in which case the presence of a certain formal connection between antecedent and consequent is an indispensable condition of the meaningfulness and truth of the implication. The concept of formal implication is not, perhaps, quite clear, but, at any rate, it is narrower than that of material implication; every meaningful and true formal implication is at the same time a meaningful and true material implication, but not vice versa.

In order to illustrate the foregoing remarks, let us consider the following four sentences:

if $2\cdot 2 = 4$, *then New York is a large city;*

if $2\cdot 2 = 5$, *then New York is a large city;*

if $2\cdot 2 = 4$, *then New York is a small city;*

if $2\cdot 2 = 5$, *then New York is a small city.*

In everyday language, these sentences would hardly be considered as meaningful, and even less as true. From the point of view of mathematical logic, on the other hand, they are all meaningful, the third sentence being false, while the remaining three are true. Thereby it is, of course, not asserted that sentences like these are particularly relevant from any viewpoint whatever, or that we apply them as premisses in our arguments.

It would be a mistake to think that the difference between everyday language and the language of logic, which has been brought to light here, is of an absolute character, and that the rules, outlined above, of the usage of the words "*if* ···, *then* ···" in common language admit of no exceptions. Actually, the usage of these words fluctuates more or less, and if we look around, we can find cases in which this usage does not comply with our rules. Let us imagine that a friend of ours is confronted with a very difficult problem and that we do not believe that he will ever solve it. We can then express our disbelief in a jocular form by saying:

if you solve this problem, I shall eat my hat.

The tendency of this utterance is quite clear. We affirm here an implication whose consequent is undoubtedly false; therefore, since we affirm the truth of the whole implication, we thereby, at the same time, affirm the falsity of the antecedent; that is to say, we express our conviction that our friend will fail to solve the problem in which he is interested. But it is also quite clear that the antecedent and the consequent of our implication are in no way connected, so that we have a typical case of a material and not of a formal implication.

The divergency in the usage of the phrase "*if* ..., *then* ..." in ordinary language and mathematical logic has been at the root of lengthy and even passionate discussions,—in which, by the way, professional logicians took only a minor part.[3] (Curiously enough, considerably less attention was paid to the analogous divergency in the case of the word "*or*".) It has been objected that the logicians, on account of their employment of the material implication, arrived at paradoxes and even plain nonsense. This has resulted in an outcry for a reform of logic to the effect of bringing about a far-reaching rapprochement between logic and ordinary language regarding the use of implication.

It would be hard to grant that these criticisms are well-founded. There is no phrase in ordinary language that has a precisely determined sense. It would scarcely be possible to find two people who would use every word with exactly the same meaning, and even in the language of a single person the meaning of the same word varies from one period of his life to another. Moreover, the meaning of words of everyday language is usually very complicated; it depends not only on the external form of the word but also on the circumstances in which it is uttered and sometimes even on subjective psychological factors. If a scientist wants to

[3] It is interesting to notice that the beginning of this discussion dates back to antiquity. It was the Greek philosopher PHILO OF MEGARA (in the 4th century B.C.) who presumably was the first in the history of logic to propagate the usage of material implication; this was in opposition to the views of his master, DIODORUS CRONUS, who proposed to use implication in a narrower sense, rather related to what is called here the formal meaning. Somewhat later (in the 3d century B.C.)—and probably under the influence of PHILO—various possible conceptions of implication were discussed by the Greek philosophers and logicians of the Stoic School (in whose writings the first beginnings of sentential calculus are to be found).

introduce a concept from everyday life into a science and to establish general laws concerning this concept, he must always make its content clearer, more precise and simpler, and free it from inessential attributes; it does not matter here whether he is a logician concerned with the phrase "*if* ..., *then* ..." or, for instance, a physicist establishing the exact meaning of the word "*metal*". In whatever way the scientist realizes his task, the usage of the term as it is established by him will deviate more or less from the practice of everyday language. If, however, he states explicitly in what meaning he decides to use the term, and if he consistently acts in conformity to this decision, nobody will be in a position to object that his procedure leads to nonsensical results.

Nevertheless, in connection with the discussions that have taken place, some logicians have undertaken attempts to reform the theory of implication. They do not, generally, deny material implication a place in logic, but they are anxious to find also a place for another concept of implication, for instance, of such a kind that the possibility of deducing the consequent from the antecedent constitutes a necessary condition for the truth of an implication; they even desire, so it seems, to place the new concept in the foreground. These attempts are of a relatively recent date, and it is too early to pass a final judgment as to their value.[4] But it appears today almost certain that the theory of material implication will surpass all other theories in simplicity, and, in any case, it must not be forgotten that logic, founded upon this simple concept, turned out to be a satisfactory basis for the most complicated and subtle mathematical reasonings.*

9. The use of implication in mathematics

The phrase "*if* ..., *then* ..." belongs to those expressions of logic which are used most frequently in other sciences and, especially, in mathematics. Mathematical theorems, particularly those of a universal character, tend to have the form of implications; the antecedent is called in mathematics the HYPOTHESIS, and the consequent is called the CONCLUSION.

[4] The first attempt of this kind was made by the contemporary American philosopher and logician C. I. LEWIS.

As a simple example of a theorem of arithmetic, having the form of an implication, we may quote the following sentence:

if x is a positive number, then 2x is a positive number

in which "*x is a positive number*" is the hypothesis, while "*2x is a positive number*" is the conclusion.

Apart from this, so to speak, classical form of mathematical theorems, there are, occasionally, different formulations, in which hypothesis and conclusion are connected in some other way than by the phrase "*if* ..., *then* ...". The theorem just mentioned, for instance, can be paraphrased in any of the following forms:

from: x is a positive number, it follows: 2x is a positive number;

the hypothesis: x is a positive number, implies (or *has as a consequence*) *the conclusion: 2x is a positive number;*

the condition: x is a positive number, is sufficient for 2x to be a positive number;

for 2x to be a positive number it is sufficient that x be a positive number;

the condition: 2x is a positive number, is necessary for x to be a positive number;

for x to be a positive number it is necessary that 2x be a positive number.

Therefore, instead of asserting a conditional sentence, one might usually just as well say that the hypothesis IMPLIES the conclusion or HAS it AS A CONSEQUENCE, or that it is a SUFFICIENT CONDITION for the conclusion; or one can express it by saying that the conclusion FOLLOWS from the hypothesis, or that it is a NECESSARY CONDITION for the latter. A logician may raise various objections against some of the formulations given above, but they are in general use in mathematics.

* The objections which might be raised here concern those of the above formulations in which any of the words "*hypothesis*", "*conclusion*", "*consequence*", "*follows*", "*implies*" occur.

In order to understand the essential points in these objections, we observe first that those formulations differ in content from the ones originally given. While in the original formulation we talk only about numbers, properties of numbers, operations upon numbers and so on—hence, about things with which mathematics is concerned—, in the formulations now under discussion we talk about hypotheses, conclusions, conditions, that is about sentences or sentential functions occurring in mathematics. It might be noted on this occasion that, in general, people do not distinguish clearly enough the terms which denote things dealt with in a given science from those which denote various kinds of expressions occurring within it. This can be observed, in particular, in the domain of mathematics, especially on the elementary level. Presumably only few are aware of the fact that such terms as "equation", "inequality", "polynomial" or "algebraic fraction", which are met at every turn in textbooks of elementary algebra, do not, strictly speaking, belong to the domain of mathematics or logic, since they do not denote things considered in this domain; equations and inequalities are certain special sentential functions, while polynomials and algebraic fractions—especially as they are treated in elementary textbooks—are particular instances of designatory functions (cf. Section 2). The confusion on this point is brought about by the fact that terms of this kind are frequently used in the formulation of mathematical theorems. This has become a very common usage, and perhaps it is not worth our while to put up a stand against it, since it does not present any particular danger; but it might be worth our while to get to recognize that, for every theorem formulated with the help of such terms, there is another formulation, logically more correct, in which those terms do not occur at all. For instance, the theorem:

the equation: $x^2 + ax + b = 0$ *has at most two roots*

can be expressed in a more correct manner as follows:

there are at most two numbers x such that $x^2 + ax + b = 0$.

Returning to the questionable formulations of an implication, we must emphasize one still more important point. In these formulations we assert that one sentence, namely the antecedent of the implication, has another—the consequent of the implication—as a consequence, or that the second follows from the first. But ordinarily when we express ourselves in this way, we have in mind that the assumption that the first sentence is true leads us, so to speak, necessarily to the same assumption concerning the second sentence (and that possibly we are even able to derive the second sentence from the first). As we already know from Section 8, however, the meaning of an implication, as it was established in contemporary logic, does not depend on whether its consequent has any such connection with its antecedent. Anyone shocked by the fact that the expression:

if $2 \cdot 2 = 4$, *then New York is a large city*

is considered in logic as a meaningful and even true sentence will find it still harder to reconcile himself with such a transformation of this phrase as:

the hypothesis that $2 \cdot 2 = 4$ *has as a consequence that New York is a large city.*

We see, thus, that the manners discussed here of formulating or transforming a conditional sentence lead to paradoxical sounding utterances and make yet more profound the discrepancies between common language and mathematical logic. It is for this reason that they repeatedly brought about various misunderstandings and have been one of the causes of those passionate and frequently sterile discussions which we mentioned above.

From the purely logical point of view we can obviously avoid all objections raised here by stating explicitly once and for all that, in using the formulations in question, we shall disregard their usual meaning and attribute to them exactly the same content as to the ordinary conditional sentence. But this would be inconvenient in another respect; for there are situations—though not in logic itself, but in a field closely related to it, namely, the methodology of deductive sciences (cf. Chapter VI)—in which we talk about sentences and the relation of consequence between them, and in which we use such terms as "*implies*" and "*follows*" in a different meaning more closely akin to the ordinary one.

It would, therefore, be better to avoid those formulations altogether, all the more since we have several formulations at our disposal which are not open to any of these objections.*

10. Equivalence of sentences

We shall consider one more expression from the field of sentential calculus. It is one which is comparatively rarely met in everyday language, namely, the phrase "*if, and only if*". If any two sentences are joined up by this phrase, the result is a compound sentence called an EQUIVALENCE. The two sentences connected in this way are referred to as the LEFT and RIGHT SIDE OF THE EQUIVALENCE. By asserting the equivalence of two sentences, it is intended to exclude the possibility that one is true and the other false; an equivalence, therefore, is true if its left and right sides are either both true or both false, and otherwise the equivalence is false.

The sense of an equivalence can also be characterized in still another way. If, in a conditional sentence, we interchange antecedent and consequent, we obtain a new sentence which, in its relation to the original sentence, is called the CONVERSE SENTENCE (or the CONVERSE OF THE GIVEN SENTENCE). Let us take, for instance, as the original sentence the implication:

(I) *if x is a positive number, then $2x$ is a positive number;*

the converse of this sentence will then be:

(II) *if $2x$ is a positive number, then x is a positive number.*

As is shown by this example, it occurs that the converse of a true sentence is true. In order to see, on the other hand, that this is not a general rule, it is sufficient to replace "$2x$" by "x^2" in (I) and (II); the sentence (I) will remain true, while the sentence (II) becomes false. If, now, it happens that two conditional sentences, of which one is the converse of the other, are both true, then the fact of their simultaneous truth can also be expressed by joining the antecedent and consequent of any one of the two sentences by the words "*if, and only if*". Thus, the above two implications—the original sentence (I) and the converse sentence (II)—may be replaced by a single sentence:

x is a positive number if, and only if, $2x$ is a positive number

(in which the two sides of the equivalence may yet be interchanged).

There are, incidentally, still a few more possible formulations which may serve to express the same idea, e.g.:

from: x is a positive number, it follows: $2x$ is a positive number, and conversely;

the conditions that x is a positive number and that $2x$ is a positive number are equivalent with each other;

the condition that x is a positive number is both necessary and sufficient for $2x$ to be a positive number;

for x to be a positive number it is necessary and sufficient that $2x$ be a positive number.

Instead of joining two sentences by the phrase "*if, and only if*", it is therefore, in general, also possible to say that the RELATION OF CONSEQUENCE holds between these two sentences IN BOTH DIRECTIONS, or that the two sentences are EQUIVALENT, or, finally, that each of the two sentences represents a NECESSARY AND SUFFICIENT CONDITION for the other.

11. The formulation of definitions and its rules

The phrase "*if, and only if*" is very frequently used in laying down DEFINITIONS, that is, conventions stipulating what meaning is to be attributed to an expression which thus far has not occurred in a certain discipline, and which may not be immediately comprehensible. Imagine, for instance, that in arithmetic the symbol "$\leqq$" has not as yet been employed but that one wants to introduce it now into the considerations (looking upon it, as usual, as an abbreviation of the expression "*is less than or equal to*"). For this purpose it is necessary to define this symbol first, that is, to explain exactly its meaning in terms of expressions which are already known and whose meanings are beyond doubt. To achieve this, we lay down the following definition,—assuming that "$>$" belongs to the symbols already known:

we say that $x \leqq y$ *if, and only if, it is not the case that* $x > y$.

The definition just formulated states the equivalence of the two sentential functions:

$$x \leqq y$$

and

$$\textit{it is not the case that} \quad x > y;$$

it may be said, therefore, that it permits the transformation of the formula "$x \leqq y$" into an equivalent expression which no longer contains the symbol "$\leqq$" but is formulated entirely in terms already comprehensible to us. The same holds for any formula obtained from "$x \leqq y$" by replacing "x" and "y" by arbitrary symbols or expressions designating numbers. The formula:

$$3 + 2 \leqq 5,$$

for instance, is equivalent with the sentence:

$$\textit{it is not the case that} \quad 3 + 2 > 5;$$

since the latter is a true assertion, so is the former. Similarly, the formula:

$$4 \leqq 2 + 1$$

is equivalent with the sentence:

$$\textit{it is not the case that} \quad 4 > 2 + 1,$$

both being false assertions. This remark applies also to more complicated sentences and sentential functions; by transforming, for instance, the sentence:

$$\textit{if} \quad x \leqq y \quad \textit{and} \quad y \leqq z, \quad \textit{then} \quad x \leqq z,$$

we obtain:

$$\textit{if it is not the case that} \quad x > y \quad \textit{and if it is not the case that} \quad y > z, \quad \textit{then it is not the case that} \quad x > z.$$

In short, by virtue of the definition given above, we are in a position to transform any simple or compound sentence containing the symbol "$\leq$" into an equivalent one no longer containing it; in other words, so to speak, to translate it into a language in which the symbol "$\leq$" does not occur. And it is this very fact which constitutes the role which definitions play within the mathematical disciplines.

If a definition is to fulfil its proper task well, certain precautionary measures have to be observed in its formulation. To this effect special rules are laid down, the so-called RULES OF DEFINITION, which specify how definitions should be constructed correctly. Since we shall not here go into an exact formulation of these rules, it may merely be remarked that, on their basis, every definition may assume the form of an equivalence; the first member of that equivalence, the DEFINIENDUM, should be a short, grammatically simple sentential function containing the constant to be defined; the second member, the DEFINIENS, may be a sentential function of an arbitrary structure, containing, however, only constants whose meaning either is immediately obvious or has been explained previously. In particular, the constant to be defined, or any expression previously defined with its help, must not occur in the definiens; otherwise the definition is incorrect, it contains an error known as a VICIOUS CIRCLE IN THE DEFINITION (just as one speaks of a VICIOUS CIRCLE IN THE PROOF, if the argument meant to establish a certain theorem is based upon that theorem itself, or upon some other theorem previously proved with its help). In order to emphasize the conventional character of a definition and to distinguish it from other statements which have the form of an equivalence, it is expedient to prefix it by words such as "*we say that*". It is easy to verify that the above definition of the symbol "$\leq$" satisfies all these conditions; it has the definiendum:

$$x \leq y,$$

whereas the definiens reads:

$$\textit{it is not the case that} \quad x > y.$$

It is worth noticing that mathematicians, in laying down definitions, prefer the words "*if*" or "*in case that*" to the phrase "*if, and only if*". If, for example, they had to formulate the definition of the symbol "$\leqq$", they would, presumably, give it the following form:

we say that $x \leqq y$, *if it is not the case that* $x > y$.

It looks as if such a definition merely states that the definiendum follows from the definiens, without emphasizing that the relation of consequence also holds in the opposite direction, and thus fails to express the equivalence of definiendum and definiens. But what we actually have here is a tacit convention to the effect that "*if*" or "*in case that*", if used to join definiendum and definiens, are to mean the same as the phrase "*if, and only if*" ordinarily does.—It may be added that the form of an equivalence is not the only form in which definitions may be laid down.

12. Laws of sentential calculus

After having come to the end of our discussion of the most important expressions of sentential calculus, we shall now try to clarify the character of the laws of this calculus.

Let us consider the following sentence:

if 1 *is a positive number and* $1 < 2$, *then* 1 *is a positive number*.

This sentence is obviously true, it contains exclusively constants belonging to the field of logic and arithmetic, and yet the idea of listing this sentence as a special theorem in a textbook of mathematics would not occur to anybody. If one reflects why this is so, one comes to the conclusion that this sentence is completely uninteresting from the standpoint of arithmetic; it fails to enrich in any way our knowledge about numbers, its truth does not at all depend upon the content of the arithmetical terms occurring in it, but merely upon the sense of the words "*and*", "*if*", "*then*". In order to make sure that this is so, let us replace in the sentence under consideration the components:

1 *is a positive number*

and

$$1 < 2$$

by any other sentences from an arbitrary field; the result is a series of sentences, each of which, like the original sentence, is true; for example:

> *if the given figure is a rhombus and if the same figure is a rectangle, then the given figure is a rhombus;*

> *if today is Sunday and the sun is shining, then today is Sunday.*

In order to express this fact in a more general form, we shall introduce the variables "*p*" and "*q*", stipulating that these symbols are not designations of numbers or any other things, but that they stand for whole sentences; variables of this kind are denoted as SENTENTIAL VARIABLES. Further, we shall replace in the sentence under consideration the phrase:

1 *is a positive number*

by "*p*" and the formula:

$$1 < 2$$

by "*q*"; in this manner we arrive at the sentential function:

if p and q, then p.

This sentential function has the property that only true sentences are obtained if arbitrary sentences are substituted for "*p*" and "*q*". This observation may be given the form of a universal statement:

For any p and q, if p and q, then p.

We have here obtained a first example of a law of sentential calculus, which will be referred to as the LAW OF SIMPLIFICATION for logical multiplication. The sentence considered above was merely a special instance of this universal law—just as, for instance, the formula:

$$2 \cdot 3 = 3 \cdot 2$$

is merely a special instance of the universal arithmetical theorem:

for arbitrary numbers x and y, $x \cdot y = y \cdot x$.

In a similar way, other laws of sentential calculus can be obtained. We give here a few examples of such laws; in their formulation we omit the universal quantifier "*for any p, q,* . . ."— in accordance with the usage mentioned in Section 3, which becomes almost a rule throughout sentential calculus.

If p, then p.

If p, then q or p.

If p implies q and q implies p, then p if, and only if, q.

If p implies q and q implies r, then p implies r.

The first of these four statements is known as the LAW OF IDENTITY, the second as the LAW OF SIMPLIFICATION for logical addition, and the fourth as the LAW OF THE HYPOTHETICAL SYLLOGISM.

Just as the arithmetical theorems of a universal character state something about the properties of arbitrary numbers, the laws of sentential calculus assert something, so one may say, about the properties of arbitrary sentences. The fact that in these laws only such variables occur as stand for quite arbitrary sentences is characteristic of sentential calculus and decisive for its great generality and the scope of its applicability.

13. Symbolism of sentential calculus; truth functions and truth tables

There exists a certain simple and general method, called METHOD OF TRUTH TABLES OR MATRICES, which enables us, in any particular case, to recognize whether a given sentence from the domain of the sentential calculus is true, and whether, therefore, it can be counted among the laws of this calculus.[5]

[5] This method originates with PEIRCE (who has already been cited at an earlier occasion; cf. footnote 2 on p. 14).

In describing this method it is convenient to apply a special symbolism. We shall replace the expressions:

not; and; or; if ..., then ...; if, and only if

by the symbols:

$$\sim; \quad \wedge; \quad \vee; \quad \rightarrow \quad ; \quad \leftrightarrow$$

respectively. The first of these symbols is to be placed in front of the expression whose negation one wants to obtain; the remaining symbols are always placed between two expressions ("$\rightarrow$" stands therefore in the place of the word "*then*", while the word "*if*" is simply omitted). From one or two simpler expressions we are, in this way, led to a more complicated expression; and if we want to use the latter for the construction of further still more complicated expressions, we enclose it in parentheses.

With the help of variables, parentheses and the constant symbols listed above (and sometimes also additional constants of a similar character which will not be discussed here), we are able to write down all sentences and sentential functions belonging to the domain of sentential calculus. Apart from the individual sentential variables the simplest sentential functions are the expressions:

$$\sim p, \quad p \wedge q, \quad p \vee q, \quad p \rightarrow q, \quad p \leftrightarrow q$$

(and other similar expressions which differ from these merely in the shape of the variables used). As an example of a compound sentential function let us consider the expression:

$$(p \vee q) \rightarrow (p \wedge r),$$

which we read, translating symbols into common language:

if p or q, then p and r.

A still more complicated expression is the law of the hypothetical syllogism given above, which now assumes the form:

$$[(p \rightarrow q) \wedge (q \rightarrow r)] \rightarrow (p \rightarrow r).$$

We can easily make sure that every sentential function occurring in our calculus is a so-called TRUTH FUNCTION. This means to say that the truth or falsehood of any sentence obtained from that function by substituting whole sentences for variables depends exclusively upon the truth or falsehood of the sentences which have been substituted. As for the simplest sentential functions "$\sim p$", "$p \wedge q$", and so on, this follows immediately from the remarks made in Sections 7, 8 and 10 concerning the meaning attributed in logic to the words "*not*", "*and*", and so on. But the same applies, likewise, to compound functions. Let us consider, for instance, the function "$(p \vee q) \rightarrow (p \wedge r)$". A sentence obtained from it by substitution is an implication, and, therefore, its truth depends on the truth of its antecedent and consequent only; the truth of the antecedent, which is a disjunction obtained from "$p \vee q$", depends only on the truth of the sentences substituted for "p" and "q", and similarly the truth of the consequent depends only on the truth of the sentences substituted for "p" and "r". Thus, finally, the truth of the whole sentence obtained from the sentential function under consideration depends exclusively on the truth of the sentences substituted for "p", "q" and "r".

In order to see quite exactly how the truth or falsity of a sentence obtained by substitution from a given sentential function depends upon the truth or falsity of the sentences substituted for variables, we construct what is called the TRUTH TABLE OR MATRIX for this function. We shall begin by giving such a table for the function "$\sim p$":

p	$\sim p$
T	F
F	T

And here is the joint truth table for the other elementary functions "$p \wedge q$", "$p \vee q$", and so on:

p	q	$p \wedge q$	$p \vee q$	$p \rightarrow q$	$p \leftrightarrow q$
T	T	T	T	T	T
F	T	F	T	T	F
T	F	F	T	F	F
F	F	F	F	T	T

The meaning of these tables becomes at once comprehensible if we take the letters "T" and "F" to be abbreviations of "true sentence" and "false sentence", respectively. In the second table, for instance, we find, in the second line below the headings "p", "q" and "$p \rightarrow q$", the letters "F", "T" and "T", respectively. We gather from that that a sentence obtained from the implication "$p \rightarrow q$" is true if we substitute any false sentence for "p" and any true sentence for "q"; this, obviously, is entirely consistent with the remarks made in Section 8.—The variables "p" and "q" occurring in the tables can, of course, be replaced by any other variables.

With the help of the two above tables, called FUNDAMENTAL TRUTH TABLES, we can construct DERIVATIVE TRUTH TABLES for any compound sentential function. The table for the function "$(p \vee q) \rightarrow (p \wedge r)$", for instance, looks as follows:

p	q	r	$p \vee q$	$p \wedge r$	$(p \vee q) \rightarrow (p \wedge r)$
T	T	T	T	T	T
F	T	T	T	F	F
T	F	T	T	T	T
F	F	T	F	F	T
T	T	F	T	F	F
F	T	F	T	F	F
T	F	F	T	F	F
F	F	F	F	F	T

In order to explain the construction of this table, let us concentrate, say, on its fifth horizontal line (below the headings). We substitute true sentences for "p" and "q" and a false sentence for "r". According to the second fundamental table, we then obtain from "$p \vee q$" a true sentence and from "$p \wedge r$" a false sentence. From the whole function "$(p \vee q) \rightarrow (p \wedge r)$" we obtain then an implication with a true antecedent and a false consequent; hence, again with the help of the second fundamental table (in which we think of "p" and "q" being for the moment replaced by "$p \vee q$" and "$p \wedge r$"), we conclude that this implication is a false sentence.

The horizontal lines of a table that consist of symbols "T" and "F" are called ROWS of the table, and the vertical lines are called COLUMNS. Each row or, rather, that part of each row which is on the left of the vertical bar represents a certain substitution of true or false sentences for the variables. When constructing the matrix of a given function, we take care to exhaust all possible ways in which a combination of symbols "T" and "F" could be correlated to the variables; and, of course, we never write in a table two rows which do not differ either in the number or in the order of the symbols "T" and "F". It can then be seen very easily that the number of rows in a table depends in a simple way on the number of different variables occurring in the function; if a function contains 1, 2, 3, $\cdots$ variables of different shape, its matrix consists of $2^1 = 2, 2^2 = 4, 2^3 = 8, \cdots$ rows. As for the number of columns, it is equal to the number of partial sentential functions of different form contained in the given function (where the whole function is also counted among its partial functions).

We are now in a position to say how it may be decided whether or not a sentence of sentential calculus is true. As we know, in sentential calculus, there is no external difference between sentences and sentential functions; the only difference consisting in the fact that the expressions considered to be sentences are always completed mentally by the universal quantifier. In order to recognize whether the given sentence is true, we treat it, for the time being, as a sentential function, and construct the truth table for it. If, in the last column of this table, no symbol "F" occurs, then every sentence obtainable from the function in question by substitution will be true, and therefore our original universal sentence (obtained from the sentential function by mentally prefixing the universal quantifier) is also true. If, however, the last column contains at least one symbol "F", our sentence is false.

Thus, for instance, we have seen that in the matrix constructed for the function "$(p \vee q) \rightarrow (p \wedge r)$" the symbol "F" occurs four times in the last column. If, therefore, we considered this expression as a sentence (that is, if we prefixed to it the words "*for any p, q and r*"), we would have a false sentence. On the other hand, it can be easily verified with the help of the method of truth tables that all the laws of sentential calculus stated in Section 12, that is, the laws of simplification, identity, and so on, are true sentences. The table for the law of simplification:

$$(p \wedge q) \rightarrow p,$$

for instance, is as follows:

p	q	$p \wedge q$	$(p \wedge q) \rightarrow p$
T	T	T	T
F	T	F	T
T	F	F	T
F	F	F	T

We give here a number of other important laws of sentential calculus whose truth can be ascertained in a similar way:

$\sim[p \wedge (\sim p)]$, $\quad p \vee (\sim p)$,

$(p \wedge p) \leftrightarrow p$, $\quad (p \vee p) \leftrightarrow p$,

$(p \wedge q) \leftrightarrow (q \wedge p)$, $\quad (p \vee q) \leftrightarrow (q \vee p)$,

$[p \wedge (q \wedge r)] \leftrightarrow [(p \wedge q) \wedge r]$, $\quad [p \vee (q \vee r)] \leftrightarrow [(p \vee q) \vee r]$.

The two laws in the first line are called the LAW OF CONTRADICTION and the LAW OF EXCLUDED MIDDLE; we next have the two LAWS OF TAUTOLOGY (for logical multiplication and addition); we then have the two COMMUTATIVE LAWS, and finally the two ASSOCIATIVE LAWS. It can easily be seen how obscure the meaning of these last two laws becomes if we try to express them in ordinary language. This exhibits very clearly the value of logical symbolism as a precise instrument for expressing more complicated thoughts.

*It occurs that the method of matrices leads us to accept sentences as true whose truth seemed to be far from obvious before the application of this method. Here are some examples of sentences of this kind:

$$p \rightarrow (q \rightarrow p),$$

$$(\sim p) \rightarrow (p \rightarrow q),$$

$$(p \rightarrow q) \vee (q \rightarrow p).$$

That these sentences are not immediately obvious is due mainly to the fact that they are a manifestation of the specific usage of implication characteristic of modern logic, namely, the usage of implication in material meaning.

These sentences assume an especially paradoxical character if, when reading them in words of common language, the implications are replaced by phrases containing "*implies*" or "*follows*", that is, if we give them, for instance, the following form:

if p is true, then p follows from any q (in other words: *a true sentence follows from every sentence*);

if p is false, then p implies any q (in other words: *a false sentence implies every sentence*);

for any p and q, either p implies q or q implies p (in other words: *at least one of any two sentences implies the other*).

In this formulation, these statements have frequently been the cause of misunderstandings and superfluous discussions. This confirms entirely the remarks made at the end of Section 9.*

14. Application of laws of sentential calculus in inference

Almost all reasonings in any scientific domain are based explicitly or implicitly upon laws of sentential calculus; we shall try to explain by means of an example in what way this happens.

Given a sentence having the form of an implication, we can, apart from its converse of which we had already spoken in Section 10, form two further sentences: the INVERSE SENTENCE (or the INVERSE OF THE GIVEN SENTENCE) and the CONTRAPOSITIVE SENTENCE. The inverse sentence is obtained by replacing both the antecedent and the consequent of the given sentence by their negations. The contrapositive is the result of interchanging the antecedent and the consequent in the inverse sentence; the contrapositive sentence is, therefore, the converse of the inverse sentence and also the inverse of the converse sentence. The converse, the inverse and the contrapositive sentences, together with the original sentence, are referred to as CONJUGATE SENTENCES. As an illustration we may consider the following conditional sentence:

(I) *if x is a positive number, then 2x is a positive number,* and form its three conjugate sentences:

if $2x$ is a positive number, then x is a positive number;

if x is not a positive number, then $2x$ is not a positive number;

if $2x$ is not a positive number, then x is not a positive number.

In this particular example, all the conjugate sentences obtained from a true sentence turn out to be likewise true. But this is not at all so in general; in order to see that it is quite possible that not only the converse sentence (as has already been mentioned in Section 10) but also the inverse sentence may be false, although the original sentence is true, it is sufficient to replace "$2x$" by "x^2" in the above sentences.

Thus it is seen that from the validity of an implication nothing definite can be inferred about the validity of the converse or the inverse sentence. The situation is quite different in the case of the fourth conjugate sentence; whenever an implication is true, the same applies to the corresponding contrapositive sentence. This fact may be confirmed by numerous examples, and it finds its expression in a general law of sentential calculus, namely the so-called LAW OF TRANSPOSITION OF OF CONTRAPOSITION.

In order to be able to formulate this law with precision, we observe that every implication may be given the schematic form:

if p, then q;

the converse, the inverse and the contrapositive sentences will then assume the forms:

if q, then p; *if not p, then not q;* *if not q, then not p.*

The law of contraposition, according to which any conditional sentence implies the corresponding contrapositive sentence, may hence be formulated as follows:

if: if p, then q, then: if not q, then not p.

In order to avoid the accumulation of the words "*if*" it is expedient to make a slight change in the formulation:

(II) *from: if p, then q, it follows that: if not q, then not p.*

We now want to show how, with the help of this law, we can, from a statement having the form of an implication—for instance, from statement (I)—derive its contrapositive statement.

(II) applies to arbitrary sentences "*p*" and "*q*", and hence remains valid if for "*p*" and "*q*" the expressions:

x is a positive number

and

2x is a positive number

are substituted. Changing, for stylistic reasons, the position of the word "*not*", we obtain:

(III) *from: if x is a positive number, then 2x is a positive number, it follows that: if 2x is not a positive number, then x is not a positive number.*

Now compare (I) and (III): (III) has the form of an implication, (I) being its hypothesis. Since the whole implication as well as its hypothesis have been acknowledged as true, the conclusion of the implication must likewise be acknowledged as true; but that is just the contrapositive statement in question:

(IV) *if 2x is not a positive number, then x is not a positive number.*

Anyone knowing the law of contraposition can, in this way, recognize the contrapositive sentence as true, provided he has previously proved the original sentence. Further, as one can easily verify, the inverse sentence is contrapositive with respect to the converse of the original sentence (that is to say, the inverse sentence can be obtained from the converse sentence by replacing antecedent and consequent by their negations and then interchanging them); for this reason, if the converse of the given sentence has been proved, the inverse sentence may likewise be considered valid. If, therefore, one has succeeded in proving two sentences—the original and its converse—a special proof for the two remaining conjugate sentences is superfluous.

It may be mentioned that several variants of the law of contraposition are known; one of them is the converse of (II):

from: if not q, then not p, it follows that: if p, then q.

This law makes it possible to derive the original sentence from the contrapositive, and the inverse from the converse sentence.

15. Rules of inference, complete proofs

We shall now consider in a little more detail the mechanism itself of the proof by means of which the sentence (IV) had been demonstrated in the preceding section. Besides the rules of definition, of which we have already spoken, we have other rules of a somewhat similar character, namely, the RULES OF INFERENCE OR RULES OF PROOF. These rules, which must not be mistaken for logical laws, amount to directions as to how sentences already known as true may be transformed so as to yield new true sentences. In the proof carried out above, two rules of demonstration have been made use of: the RULE OF SUBSTITUTION and the RULE OF DETACHMENT (also known as the MODUS PONENS RULE).

The content of the rule of substitution is as follows. If a sentence of a universal character, that has already been accepted as true, contains sentential variables, and if these variables are replaced by other sentential variables or by sentential functions or by sentences—always substituting equal expressions for equal variables throughout—, then the sentence obtained in this way may also be recognized as true. It was by applying this very rule that we obtained the sentence (III) from sentence (II). It should be emphasized that the rule of substitution may also be applied to other kinds of variables, for example, to the variables "x", "y", ··· designating numbers: in place of these variables, any symbols or expressions denoting numbers may be substituted.

*The formulation of the rule of substitution as given here is not quite precise. This rule refers to such sentences as are composed of a universal quantifier and a sentential function, the latter containing variables bound by the universal quantifier. When one wants to apply the rule of substitution, one omits the quantifier and substitutes for the variables previously bound by this quantifier either other variables or whole expressions (e.g., sentential functions or sentences for the variables "p", "q", "r", ··· , and expressions denoting numbers for the variables "x", "y", "z", ···);

any other bound variables which may occur in the sentential function remain unaltered, and one sees to it that no variables of the same form as these occur in the substituted expressions; if necessary a universal quantifier is set in front of the expression obtained in this way in order to turn it into a sentence. For instance, by applying the rule of substitution to the sentence:

for any number x there is a number y such that $x + y = 5$,

the following sentence can be obtained:

there is a number y such that $3 + y = 5$,

but also the sentence:

for any number z there is a number y such that $z^2 + y = 5$;

thus, in this case, one substitutes only for "x" and leaves "y" unaltered. We must not, however, substitute for "x" any expression containing "y"; for, although our original sentence was true, we might in this way arrive at a false sentence. For instance, by substituting "$3 - y$", we should obtain:

there is a number y such that $(3 - y) + y = 5$.*

The rule of detachment states that, if two sentences are accepted as true, of which one has the form of an implication while the other is the antecedent of this implication, then that sentence may also be recognized as true which forms the consequent of the implication. (We "detach" thus, so to speak, the antecedent from the whole implication.) By means of this rule, the sentence (IV) had been derived from the sentences (III) and (I).

It can be seen that from this that, in the proof of the sentence (IV) as carried out above, each step consisted in applying a rule of inference to sentences which were previously accepted or recognized as true. A proof of this kind will be called COMPLETE. A little more precisely a complete proof may also be characterized as follows. It consists in the construction of a chain of sentences with these properties: the initial members are sentences which were already previously accepted as true; every subsequent member is obtainable from preceding ones by applying a rule of inference; and finally the last member is the sentence to be proved.

It should be observed what an extremely elementary form—from the psychological point of view—all mathematical reasonings assume, due to the knowledge and application of the laws of logic and the rules of inference; complicated mental processes are entirely reducible to such simple activities as the attentive observation of statements previously accepted as true, the perception of structural, purely external, connections among these statements, and the execution of mechanical transformations as prescribed by the rules of inference. It is obvious that, in view of such a procedure, the possibility of committing mistakes in a proof is reduced to a minimum.

Exercises

1. Give examples of specifically mathematical expressions from the fields of arithmetic and geometry.

2. Differentiate in the following two sentences between the specifically mathematical expressions and those belonging to the domain of logic:

(a) *for any numbers x and y, if* $x > 0$ *and* $y < 0$, *then there is a number z such that* $z < 0$ *and* $x = y \cdot z$;

(b) *for any points A and B there is a point C which lies between A and B and is the same distance from A as from B.*

3. Form the conjunction of the negations of the following sentential functions:

$$x < 3$$

and

$$x > 3.$$

What number satisfies this conjunction?

4. In which of its two meanings does the word "*or*" occur in the following sentences:

(a) *two ways were open to him: to betray his country or to die;*

(b) *if I earn a lot of money or win the sweepstake, I shall go on a long journey.*

Give further examples in which the word *"or"* is used in its first or in its second meaning.

*5. Consider the following conditional sentences:

(a) *if today is Monday, then tomorrow is Tuesday;*

(b) *if today is Monday, then tomorrow is Saturday;*

(c) *if today is Monday, then the 25th of December is Christmas day;*

(d) *if wishes were horses, beggars could ride;*

(e) *if a number x is divisible by* 2 *and by* 6, *then it is divisible by* 12;

(f) *if* 18 *is divisible by* 3 *and by* 4, *then* 18 *is divisible by* 6.

Which of the above implications are true and which are false from the point of view of mathematical logic? In which cases does the question of meaningfulness and of truth or falsity raise any doubts from the standpoint of ordinary language? Direct special attention to the sentence (b) and examine the question of its truth as dependent on the day of the week on which it was uttered.

6. Put the following theorems into the form of ordinary conditional sentences:

(a) *for a triangle to be equilateral, it is sufficient that the angles of the triangle be congruent;*

(b) *the condition: x is divisible by* 3, *is necessary for x to be divisible by* 6.

Give further paraphrases of these two sentences.

7. Is the condition:

$$x \cdot y > 4$$

necessary or sufficient for the validity of:

$$x > 2 \quad \textit{and} \quad y > 2 \quad ?$$

8. Give alternative formulations for the following sentences:

(a) *x is divisible by 10 if, and only if, x is divisible both by 2 and by 5;*

(b) *for a quadrangle to be a parallelogram it is necessary and sufficient that the point of intersection of its diagonals be at the same time the midpoint of each diagonal.*

Give further examples of theorems from the fields of arithmetic and geometry that have the form of equivalences.

9. Which of the following sentences are true:

(a) *a triangle is isosceles if, and only if, all the altitudes of the triangle are congruent;*

(b) *the fact that $x \neq 0$ is necessary and sufficient for x^2 to be a positive number;*

(c) *the fact that a quadrangle is a square implies that all its angles are right angles, and conversely;*

(d) *for x to be divisible by 8 it is necessary and sufficient that x be divisible both by 4 and by 2* ?

10. Assuming the terms "*natural number*" and "*product*" (or "*quotient*", respectively) to be known already, construct the definition of the term "*divisible*", giving it the form of an equivalence:

we say that x is divisible by y if, and only if, . . .

Likewise formulate the definition of the term "*parallel*"; what terms (from the domain of geometry) have to be presupposed for this purpose?

11. Translate the following symbolic expressions into ordinary language:

(a) $[(\sim p) \rightarrow p] \rightarrow p$,

(b) $[(\sim p) \vee q] \leftrightarrow (p \rightarrow q)$,

(c) $[\sim(p \vee q)] \leftrightarrow (p \rightarrow q)$,

(d) $(\sim p) \vee [q \leftrightarrow (p \rightarrow q)]$.

Direct special attention to the difficulty in distinguishing in ordinary language the three last expressions.

12. Formulate the following expressions in logical symbolism:

(a) *if not p or not q, then it is not the case that p or q;*

(b) *if p implies that q implies r, then p and q together imply r;*

(c) *if r follows from p and if r follows from q, then r follows from p or q.*

13. Construct truth tables for all sentential functions given in Exercises 11 and 12. Assume that we interpret these functions as sentences (what does this mean?), and determine which of these sentences are true and which are false.

14. Verify by the method of truth tables that the following sentences are true:

(a) $[\backsim(\backsim p)] \leftrightarrow p,$

(b) $[\backsim(p \wedge q)] \leftrightarrow [(\backsim p) \vee (\backsim q)],$
$[\backsim(p \vee q)] \leftrightarrow [(\backsim p) \wedge (\backsim q)],$

(c) $[p \wedge (q \vee r)] \leftrightarrow [(p \wedge q) \vee (p \wedge r)],$
$[p \vee (q \wedge r)] \leftrightarrow [(p \vee q) \wedge (p \vee r)].$

Sentence (a) is the LAW OF DOUBLE NEGATION, sentences (b) are called DE MORGAN'S LAWS[6], and sentences (c) are the DISTRIBUTIVE LAWS (for logical multiplication with respect to addition and for logical addition with respect to multiplication).

15. For each of the following sentences, state the three corresponding conjugate sentences (the converse, the inverse, and the contrapositive sentence):

(a) *the fact that x is a positive number implies that −x is a negative number;*

(b) *if a quadrangle is a rectangle, then a circle can be circumscribed about it.*

Which of the conjugate sentences are true?

Give an example of four conjugate sentences which are all false.

[6] These laws were given by A. DE MORGAN (1806–1878), an eminent English logician.

16. Explain the following fact on the basis of the truth table for the function "$p \leftrightarrow q$": if in any sentence some of its parts which are themselves sentences (or sentential functions) are replaced by equivalent sentences, then the whole new sentence obtained in this way is equivalent to the original sentence. Some of our statements and remarks in Section 10 were dependent on this fact; indicate where this was the case.

17. Consider the following two sentences:

(a) *from: if p, then q, it follows that: if q, then p;*

(b) *from: if p, then q, it follows that: if not p, then not q.*

Suppose these sentences were logical laws, would it be possible to apply them in mathematical proofs in an analogous way to the law of contraposition (cf. Section 14)? What conjugate sentences would it be possible to derive from a given asserted implication? Consequently, can our supposition be maintained that the sentences (a) and (b) are true?

18. Confirm the conclusion which has been reached in Exercise 17 by applying the method of truth tables to the sentences (a) and (b).

19. Consider the following two statements:

the fact that yesterday was Monday implies that today is Tuesday;

the fact that today is Tuesday implies that tomorrow will be Wednesday.

What statement may be deduced from them in accordance with the law of the hypothetical syllogism (cf. Section 12)?

*20. Carry out the complete proof of the statement obtained in the preceding exercise; use the statements and law of the hypothetical syllogism mentioned there, and apply—in addition to the rule of substitution and the rule of detachment—the following rule of inference: if any two sentences are accepted as true, then their conjunction may be recognized as true.

· III ·

ON THE THEORY OF IDENTITY

16. Logical concepts outside sentential calculus; concept of identity

Sentential calculus, to which the preceding chapter was devoted, forms merely a part of logic. It constitutes undoubtedly the most fundamental part,—at least inasmuch as one makes use of the terms and laws of this calculus in the definition of terms and the formulation and demonstration of logical laws that do not belong to sentential calculus. Sentential calculus in itself, however, does not form a sufficient basis for the foundation of other sciences and, in particular, not of mathematics; various concepts from other parts of logic are constantly encountered in mathematical definitions, theorems and proofs. Some of them will be discussed in the present and in the following two chapters.

Among the logical concepts not belonging to sentential calculus, the concept of IDENTITY or EQUALITY is probably the one which has the greatest importance. It occurs in phrases such as:

x is identical with y,

x is the same as y,

x equals y.

To all three of these expressions the same meaning is ascribed; for the sake of brevity, they will be replaced by the symbolic expression:

$$x = y.$$

Instead of writing:

x is not identical with y

or:

x is different from y

we employ the formula:

$$x \neq y.$$

The general laws involving the above expressions constitute a part of logic which may be called the THEORY OF IDENTITY.

17. Fundamental laws of the theory of identity

Among the logical laws concerning the concept of identity the most fundamental is the following:

I. $x = y$ *if, and only if, x has every property which y has, and y has every property which x has.*

We could also say more simply:

$x = y$ *if, and only if, x and y have every property in common.*

Other and perhaps more apparent, though less correct, formulations of the same law are known, for instance:

$x = y$ *if, and only if, everything that may be said about any one of the things x or y may also be said about the other.*

Law I was first stated by LEIBNIZ[1] (although in somewhat different terms) and hence may be called LEIBNIZ'S LAW. It has the form of an equivalence, and enables us to replace the formula:

$$x = y,$$

which is the left side of the equivalence, by its right side, that is by an expression no longer containing the symbol of identity. With respect to its form this law may, therefore, be considered as the definition of the symbol "=", and so it was considered by LEIBNIZ himself. (Of course, to regard LEIBNIZ's law here as a definition would make sense only if the meaning of the symbol "=" seemed to us less evident than that of the expressions on the right side of the law, such as "*x has every property which y has*"; cf. Section 11).

[1] Cf. footnote 1 on p. 19.

As a consequence of LEIBNIZ's law we have the following rule which is of great practical importance: If, in a certain context, a formula having the form of an equation, e.g.:

$$x = y,$$

has been assumed or proved, then it is permissible to replace, in any formula or sentence occurring in this context, the left side of the equation by its right side, e.g. "x" by "y", and conversely. It is understood that, should "x" occur at several places in a formula, it may at some places be left unchanged and at others replaced by "y"; there is, thus, an essential difference between the rule now under consideration and the rule of substitution discussed in Section 15 which does not permit such a partial replacement of one symbol by another.

From LEIBNIZ's law we can derive a number of other laws belonging to the theory of identity, that are frequently applied in various considerations and especially in mathematical proofs. The most important among these will be listed here, together with a sketch of their proofs, in order to exhibit by way of concrete examples that there is no essential difference between reasonings in the field of logic, and those in the field of mathematics.

II. *Everything is equal to itself:* $x = x$.

PROOF. Substitute, in LEIBNIZ's law, "x" for "y"; we obtain:

$x = x$ *if, and only if, x has every property which x has, and x has every property which x has.*

We can, of course, simplify this sentence by omitting its last part "*and x has* ..." (this follows directly from the law of tautology stated in Section 12). The sentence assumes then the following form:

$x = x$ *if, and only if, x has every property which x has.*

Obviously, the right side of this equivalence is always satisfied (for, according to the law of identity in Section 12, if x has a certain property, it has this property). Hence the left side of the equivalence must also be satisfied; in other words, we always have:

$$x = x,$$

which was to be proved.

III. *If* $x = y$, *then* $y = x$.

PROOF. By substituting, in LEIBNIZ's law, "x" for "y" and "y" for "x", we obtain:

$y = x$ *if, and only if, y has every property which x has, and x has every property which y has.*

Let us compare this sentence with the original formulation of LEIBNIZ's law. We have here two equivalences, the right sides of which are conjunctions differing only in the order of their members. Hence the right sides are equivalent (cf. the commutative law of logical multiplication in Section 13), and the left sides, that is, the formulas:

$$x = y \quad \text{and} \quad y = x$$

must, therefore, be also equivalent. A fortiori we may assert that the second of these formulas follows from the first, as it is stated in our law.

IV. *If* $x = y$ *and* $y = z$, *then* $x = z$.

PROOF. By hypothesis, the two formulas:

(1) $$x = y$$

and

(2) $$y = z$$

are assumed valid. According to LEIBNIZ's law, it follows from formula (2) that everything that may be said about y may also be said about z. Hence we may replace the variable "y" by "z" in formula (1), and we obtain the desired formula:

$$x = z.$$

V. *If* $x = z$ *and* $y = z$, *then* $x = y$; in other words, *two things equal to the same thing are equal to each other.*

This law can be proved in a way quite analogous to the preceding (it can also be deduced from Laws III and IV, without using LEIBNIZ's law).

Laws II, III and IV are called the LAWS OF REFLEXIVITY, OF SYMMETRY, and OF TRANSITIVITY for the relation of identity.

18. Identity of things and identity of their designations; use of quotation marks

*Although the meaning of such expressions as:

$$x = y \quad \text{or} \quad x \neq y$$

seems to be evident, these expressions are sometimes misunderstood. It seems obvious, for instance, that the formula:

$$3 = 2 + 1$$

is a true assertion, and yet some people are somewhat doubtful as to its truth. In their opinion, this formula appears to state that the symbols "3" and "2 + 1" are identical, which is obviously false since these symbols have entirely different shapes, and, therefore, it is not true that everything that may be said about one of these symbols may be said about the other (for instance, the first symbol is a single sign, while the second is not).

In order to avoid doubts of this kind, it is well to make clear to oneself a very general and important principle upon which the useful employability of any language is dependent. According to this principle, whenever, in a sentence, we wish to say something about a certain thing, we have to use, in this sentence, not the thing itself but its name or designation.

The application of this principle gives no cause for doubt as long as the thing talked about is not a word, a symbol or, more generally, an expression of a language. Let us imagine, for example, that we have a small blue stone in front of us, and that we state the following sentence:

this stone is blue.

To none, presumably, would it occur in this case, to replace in this sentence the words "*this stone*" which together constitute the designation of the thing by the thing itself, that is to say, to blot or cut these words out and to place in their stead the stone. For, in doing so, we would arrive at a whole consisting in part of a stone and in part of words, and thus at something which would not be a linguistic expression, and far less a true sentence.

This principle is, however, frequently violated if the thing talked about happens to be a word or a symbol. And yet the application of the principle is indispensable also in this case; for, otherwise, we would arrive at a whole which, though being a linguistic expression, would fail to express the thought intended by us, and very often might even be a meaningless aggregate of words. Let us consider, for example, the following two words:

well, Mary.

Clearly, the first consists of four letters, and the second is a proper name. But let us imagine that we would express these thoughts, which are quite correct, in the following manner:

(I) *well consists of four letters;*

(II) *Mary is a proper name;*

we would then, in talking about words, be using the words themselves and not their names. And if we examine the expressions (I) and (II) more closely, we must admit that the first is not a sentence at all since the subject can only be a noun and not an adverb, while the second might be considered a meaningful sentence, but, at any rate, a false one since no woman is a proper name.

In order to avoid these difficulties, we might assume that the words "*well*" and "*Mary*" occur in such contexts as (I) and (II) in a meaning distinct from the usual one, and that they here function as their own names. In generalization of this viewpoint, we should have to admit that any word may, at times, function as its own name; to use the terminology of medieval logic, we may say that, in a case like this, the word is used in SUPPOSITIO MATERIALIS, as opposed to its use in SUPPOSITIO FORMALIS, that is, in its ordinary meaning. As a consequence, every word of common or scientific language would possess at least two different meanings, and one would not have to look far for examples of situations in which serious doubts might arise as to which meaning was intended. With this consequence we do not wish to reconcile ourselves, and therefore we will make it a rule that every expression should differ (at least in writing) from its name.

The problem arises as to how we can set about to form names of words and expressions. There are various devices to this effect. The simplest one among them is based on the convention of forming the name of an expression by placing it between quotation marks. On the basis of this agreement, the thoughts tentatively expressed in (I) and (II) can now be stated correctly and without ambiguity, thus:

(I′) *"well" consists of four letters;*

(II′) *"Mary" is a proper name.*

In the light of these remarks all possible doubts as to the meaning and the truth of such formulas as:

$$3 = 2 + 1$$

are dispelled. This formula contains symbols designating certain numbers, but it does not contain the names of any such symbols. Therefore this formula states something about numbers and not about symbols designating numbers; the numbers 3 and 2 + 1 are obviously equal, so that the formula is a true assertion. We may, admittedly, replace this formula by an equivalent sentence which is about symbols, namely, we may say that the symbols "3" and "2 + 1" designate the same number. But this by no means implies that the symbols themselves are identical; for it is well known that the same thing—and, in particular, the same number—can have many different designations. The symbols "3" and "2 + 1" are, no doubt, different, and this fact can be expressed in the form of a new formula:

$$\text{“3”} \neq \text{“2 + 1”}$$

which, of course, in no way contradicts the formula previously stated.[2*]

[2] The convention concerning the use of quotation marks has been adhered to in this book pretty consistently. We deviate from it only in rare cases, by way of a concession to traditional usage. For instance, we state formulas and sentences without quotation marks, if they are printed

19. Equality in arithmetic and geometry, and its relation to logical identity

We here consider the notion of arithmetical equality among numbers consistently as a special case of the general concept of logical identity. It must be added, however, that there are mathematicians who—as opposed to the standpoint adopted here—do not identify the symbol "=" occurring in arithmetic with the symbol of logical identity; they do not consider equal numbers to be necessarily identical, and therefore look upon the notion of equality among numbers as a specifically arithmetical concept. In this connection, those mathematicians reject LEIBNIZ's law in its general form, and merely recognize some of its consequences which are of a less general character and count them among the specifically mathematical theorems. Among these consequences there are Laws II to V of Section 17, as well as theorems to the effect that, whenever $x = y$ and x satisfies some formula built up of arithmetical symbols only, then y satisfies the same formula; thus, for instance, the theorem:

$$\textit{if} \quad x = y \quad \textit{and} \quad x < z, \quad \textit{then} \quad y < z.$$

In our opinion, this point of view can claim no particular theoretical advantages, while, in practice, it entails considerable complications in the presentation of the system of arithmetic. For one rejects here the general rule which allows us—on the assumption that an equation holds—to replace everywhere the left side of the equation by its right side; since, however, such a replacement is indispensable in various arguments, it becomes necessary to give a special proof that this replacement is permissible in each particular case in which it is applied.

---

displayed in a special line or if they occur in the formulation of mathematical or logical theorems; and we do not put quotation marks about expressions which are preceded by such phrases as "is called", "is known as", and so on. But other precautionary measures are taken in these cases; the expression in question is often preceded by a colon and it is usually printed in a different kind of print (small capitals or italics). It should be observed that, in everyday language, quotation marks are used also in cases not covered by the above convention; and examples of this type can be found in this book, too.

To illustrate this situation by an example, let us consider any system of equations in two variables, for instance:

$$x = y^2,$$
$$x^2 + y^2 = 2x - 3y + 18.$$

If one wants to solve this system of equations by means of the so-called method of substitution, one has to form a new system of equations obtained by leaving the first equation unchanged and replacing in the second equation "x" by "y^2" throughout. And the question arises whether this transformation is permissible, that is, whether the new system is equivalent to the old. The answer is undoubtedly in the affirmative, no matter what conception of the notion of equality among numbers is adopted. But if the symbol "=" is understood to designate logical identity, and if LEIBNIZ's law is assumed, the answer is obvious; the assumption:

$$x = y^2$$

permits us to replace "x" everywhere by "y^2", and vice versa. Otherwise it would first be necessary to give reasons for the affirmative answer, and although this justification would not meet with any essential difficulties, it would at any rate be rather long and tedious.

As to the notion of equality in geometry, the situation is entirely different. If two geometrical figures, such as two line segments, or two angles, or two polygons, are said to be equal or congruent, it is in general not intended to assert their identity. One merely wishes to state that the two figures have the same size and shape, in other words—to use a figurative if not quite correct mode of expression—, that they would exactly cover one another if one were placed on top of the other. Thus, for example, a triangle is capable of having two, or even three, equal sides, and yet these sides are certainly not identical. There are also cases, on the other hand, in which it is not a question of the geometrical equality of two figures, but of their logical identity; for instance, in an isosceles triangle, the altitude upon and the median to the base are not only geometrically equal, but they are simply one and the same line segment. Therefore, in order to avoid any confusion, it would be recommendable consistently to avoid the term "equality" in all those cases where it is not a question of logical identity, and to speak of geometrically equal figures rather as of congruent figures, replacing—as it is often done anyhow—the symbol "$=$" by a different one, such as "$\cong$".

20. Numerical quantifiers

With the help of the concept of identity it is possible to give a precise meaning to certain phrases which, both in their content and their function, are closely related to the universal and existential quantifiers and are also counted among the operators, but which are of a more special character. They are expressions such as:

> *there is at least one*, or *at most one*, or *exactly one*, *thing x such that* ⋯ ,

> *there are at least two*, or *at most two*, or *exactly two*, *things x such that* ⋯ ,

and so on; they might be called NUMERICAL QUANTIFIERS. Apparently, specifically mathematical terms occur in these phrases, namely, the numerals "*one*", "*two*", and so on. A more exact analysis shows, however, that the content of those phrases (if considered as a whole) is of a purely logical nature. Thus, in the expression:

> *there is at least one thing satisfying the given condition*

the words "*at least one*" may simply be replaced by the article "*a*" without altering the meaning. The expression:

> *there is at most one thing satisfying the given condition*

means the same as:

> *for any x and y, if x satisfies the given condition and if y satisfies the given condition, then* $x = y$.

The sentence:

there is exactly one thing satisfying the given condition

is equivalent with the conjunction of the two sentences just given:

there is at least one thing satisfying the given condition, and at the same time there is at most one thing satisfying the given condition.

To the expression:

there are at least two things satisfying the given condition

we give the following meaning:

there are x and y, such that both x and y satisfy the given condition and $x \neq y$;

it is, therefore, equivalent to the negation of:

there is at most one thing satisfying the given condition.

Analogously we explain the meanings of other expressions of this category.

For the purpose of illustration, a few true sentences of arithmetic may be listed here in which numerical quantifiers appear:

there is exactly one number x, such that $x + 2 = 5$;

there are exactly two numbers y, such that $y^2 = 4$;

there are at least two numbers z, such that $z + 2 < 6$.

That part of logic, in which general laws involving quantifiers are laid down, is known as the THEORY OF APPARENT VARIABLES or the FUNCTIONAL CALCULUS, although it really ought to be called the CALCULUS OF QUANTIFIERS. Hitherto this theory has primarily concerned itself with the universal and existential quantifiers, while the numerical quantifiers have been largely neglected.

Exercises

1. Prove Law V of Section 17, using exclusively Laws III and IV, thus without the use of LEIBNIZ's law.

Hint: In Law V the formulas:

$$x = z \quad \text{and} \quad y = z$$

are assumed valid by hypothesis. By virtue of Law III, interchange the variables in the second of these formulas, and then apply Law IV.

2. Prove the following law:

if $x = y$, $y = z$ *and* $z = t$, *then* $x = t$,

using exclusively Law IV of Section 17.

3. Are the sentences true which are obtained by replacing in Laws III and IV of Section 17 the symbol "$=$" by "$\neq$" throughout?

*4. On the basis of the convention stated in Section 18 concerning the use of quotation marks, determine which of the following expressions are true sentences:

(a) 0 *is an integer,*

(b) 0 *is a cipher having an oval shape,*

(c) "0" *is an integer,*

(d) "0" *is a cipher having an oval shape,*

(e) $1.5 = \frac{3}{2}$,

(f) "1.5" = "$\frac{3}{2}$",

(g) $2 + 2 \neq 5$,

(h) "2 + 2" $\neq$ "5".

*5. In order to form the name of a word we put that word in quotation marks; in order to form the name of this name we put, in turn, the name of this word in quotation marks, and thus the word itself in double quotation marks. Hence, of the following three expressions:

John, "*John*", " "*John*" ",

the second is the name of the first, and the third is the name of the second. Substitute in turn the three expressions above for

"x" in the following sentential functions, and determine which of the twelve sentences obtained are true:

(a) *x is a man,*

(b) *x is a name of a man,*

(c) *x is an expression,*

(d) *x is an expression containing quotation marks.*

*6. In Section 9 we gave various formulations of conditional sentences met with in mathematical books. Attention was also called to the fact that in some of these formulations we talk, not about numbers or properties of numbers, and so on, but about expressions (e.g., sentences and sentential functions). It follows from remarks made in Section 18 that these latter formulations call for the use of quotation marks. Indicate the formulations, and the exact place in them, in which quotation marks are required.

*7. On the basis of the general principle concerning the use of names of things in sentences stating something about these things, we may now subject the last sentence but one of Section 12 ("Just as the arithmetical theorems ...") to some criticism. We know that variables occurring in arithmetic stand for names of numbers. Do the variables occurring in sentential calculus stand for names of sentences or for sentences themselves? May we, therefore, say, if we want to be exact, that the laws of this calculus assert something about the sentences and their properties?

8. Consider a triangle with sides a, b and c. Let h_a, h_b and h_c be the altitudes upon the sides a, b and c; similarly, let m_a, m_b and m_c be the medians and s_a, s_b and s_c the bisectors of the angles of the triangle.

Assuming the triangle to be isosceles (with a as the base, and b and c as the sides of equal length), which of the twelve segments named are congruent (i.e., equal in the geometrical sense), and which are identical? Express the answer in the form of formulas, using the symbol "$\cong$" to designate congruence, and the symbol "$=$" to designate identity.

Solve the same problem under the assumption that the triangle is equilateral.

9. Explain the meaning of the following expressions:

(a) *there are at most two things satisfying the given condition;*

(b) *there are exactly two things satisfying the given condition.*

10. Determine which of the following sentences are true:

(a) *there is exactly one number x such that* $x + 3 = 7 - x$;

(b) *there are exactly two numbers x such that* $x^2 + 4 = 4x$;

(c) *there are at most two numbers y such that* $y + 5 < 11 - 2y$;

(d) *there are at least three numbers z such that* $z^2 < 2z$;

(e) *for any number x there is exactly one number y such that* $x + y = 2$;

(f) *for any number x there is exactly one number y such that* $x \cdot y = 3$.

11. How can one make use of numerical quantifiers in order to express the fact that the equation:

$$x^2 - 5x + 6 = 0$$

has two roots?

12. What numbers x satisfy the sentential function:

there are exactly two numbers y such that $x = y^2$?

Differentiate in this function between free and bound variables. *Do numerical quantifiers bind variables?*

· IV ·

ON THE THEORY OF CLASSES

21. Classes and their elements

Apart from separate individual things, which we shall also, for short, call INDIVIDUALS, logic is concerned with CLASSES of things; in everyday life as well as in mathematics, classes are more often referred to as SETS. Arithmetic, for instance, frequently deals with sets of numbers, and in geometry our interest attaches itself not so much to single points as to point sets (namely, to geometrical configurations). Classes of individuals are called CLASSES OF THE FIRST ORDER. Comparatively more rarely we also meet in our investigations with CLASSES OF THE SECOND ORDER, that is, with classes which consist, not of individuals, but of classes of the first order. Sometimes even CLASSES OF THE THIRD, FOURTH, ... ORDERS have to be dealt with. Here we shall be concerned almost exclusively with classes of the first order, and only exceptionally—as in Section 26—we shall have to deal with classes of the second order; our considerations can, however, be applied with practically no changes to classes of any order.

In order to differentiate between individuals and classes (and also between classes of different orders), we employ as variables letters of different shape and belonging to different alphabets. It is customary to designate individual things such as numbers, and classes of such things, by the small and capital letters of the English alphabet, respectively. In elementary geometry the opposite notation is the accepted one, capital letters designating points and small letters (of the English or Greek alphabets) designating point sets.

That part of logic in which the class concept and its general properties are examined is called the THEORY OF CLASSES; sometimes this theory is also treated as an independent mathematical discipline under the name of the GENERAL THEORY OF SETS.[1]

[1] The beginnings of the theory of classes—or, to be more exact, of that part of this theory which we shall denote as the calculus of classes below—

Of fundamental character in the theory of classes are such phrases as:

the thing x is an element (or *a member*) *of the class K,*

the thing x belongs to the class K,

the class K contains the thing x as an element (or *a member*);

we consider these expressions as having the same meaning and, for the sake of brevity, replace them by the formula:

$$x \in K.$$

Thus, if **I** is the set of all integers, the numbers 1, 2, 3, ··· are its elements, whereas the numbers $\frac{2}{3}$, $2\frac{1}{2}$, ··· do not belong to the set; hence, the formulas:

$$1 \in \mathbf{I}, \quad 2 \in \mathbf{I}, \quad 3 \in \mathbf{I}, \cdots$$

are true, while the formulas:

$$\tfrac{2}{3} \in \mathbf{I}, \quad 2\tfrac{1}{2} \in \mathbf{I}, \cdots$$

are false.

22. Classes and sentential functions with one free variable

We consider a sentential function with one free variable, for instance:

$$x > 0$$

If we prefix the words:

(I) *the set of all numbers x such that*

---

are already found in G. BOOLE (cf. footnote 1 on p. 19). The actual creator of the general theory of sets as an independent mathematical discipline was the great German mathematician G. CANTOR (1845–1918); we are indebted to him, in particular, for the analysis of such concepts as equality in power, cardinal number, infinity and order, which will be discussed in the course of the present and the next chapters.—CANTOR'S set theory is one of those mathematical disciplines which are in a state of especially intensive development. Its ideas and lines of thought have penetrated into almost all branches of mathematics and have exerted everywhere a most stimulating and fertilizing influence.

to that function, we obtain the expression:

$$\textit{the set of all numbers } x \textit{ such that} \quad x > 0.$$

This expression designates a well-determined set, namely the set of all positive numbers; it is the set having as its elements those, and only those, numbers which satisfy the given function. If we denote this set by the symbol "P", our function becomes equivalent to:

$$x \in \mathrm{P}.$$

We may apply an analogous procedure to any other sentential function. In arithmetic, we can obtain in this way various sets of numbers, for instance the set of all negative numbers or the set of all numbers which are greater than 2 and less than 5 (that is which satisfy the function "$x > 2$ *and* $x < 5$"). This procedure plays also an important role in geometry, especially in defining new kinds of geometrical configurations; the surface of a sphere is defined, for instance, as the set of all points of the space which have a definite distance from a given point. It is customary in geometry to replace the words "*the set of all points*" by "*the locus of the points.*"

We will now put the above remarks in a general form. It is assumed in logic that, to every sentential function containing just one free variable, say "x", there is exactly one corresponding class having as its elements those, and only those, things x which satisfy the given function. We obtain a designation for that class by putting in front of the sentential function the following phrase, which belongs to the fundamental expressions of the theory of classes:

(II) *the class of all things x such that.*

If we denote further the class in question by a simple symbol, say "C", the formula:

$$x \in \mathrm{C}$$

will—for any x—be equivalent to the original sentential function.

Hence it is seen that any sentential function containing "x" as the only free variable can be transformed into an equivalent function of the form:

$$x \in K,$$

where in place of "K" we have a constant denoting a class; one may, therefore, consider the latter formula as the most general form of a sentential function with one free variable.

The phrases (I) and (II) are sometimes replaced by symbolic expressions; we can, for instance, agree to use the following symbol for this purpose:

$$\underset{x}{\mathsf{C}}.$$

*Let us now consider the following expression:

1 belongs to the set of all numbers x such that $x > 0$,

which can also be written in symbols only:

$$1 \in \underset{x}{\mathsf{C}}(x > 0).$$

This expression is obviously a sentence, and even a true sentence; it expresses, in a more complicated form, the same thought as the simple formula:

$$1 > 0$$

Consequently, this expression cannot contain any free variable, and the variable "x" occurring in it must be a bound variable. Since, on the other hand, we do not find in the above expression any quantifiers, we arrive at the conclusion that such phrases as (I) or (II) function like quantifiers, that is, they bind variables, and must, therefore, be counted among the operators (cf. Section 4).

It should be added that we frequently prefix an operator like (I) or (II) to sentential functions which contain—besides "x"—other free variables (this occurs in nearly all cases in which such operators are applied in geometry). The expressions thus obtained, for instance:

the set of all numbers x such that $x > y$

do not designate, however, any definite class; they are designatory functions in the meaning established in Section 2, that is, they become designations of classes if we replace in them free variables (but not "x") by suitable constants, for instance, "y" by "0" in the example just given.*

It is frequently said of a sentential function with one free variable that it expresses a certain property of things,—a property possessed by those, and only those, things which satisfy the sentential function (the sentential function "*x is divisible by 2*", for example, expresses a certain property of the number x, namely, divisibility by 2, or the property of being even). The class corresponding to this function contains as its elements all things possessing the given property, and no others. In this manner it is possible to correlate a uniquely determined class with every property of things. And also, conversely, with every class there is correlated a property possessed exclusively by the elements of that class, namely, the property of belonging to that class. It is, accordingly, in the opinion of numerous logicians, unnecessary to distinguish at all between the two concepts of a class and of a property; in other words, a special "theory of properties" is dispensable,—the theory of classes being perfectly sufficient.

As an application of these remarks we shall give a new formulation of LEIBNIZ's law. The original one (in Section 17) contained the term "*property*"; in the following, entirely equivalent, formulation we employ the term "*class*" instead:.

$x = y$ *if, and only if, every class which contains any one of the things x and y as an element also contains the other as an element.*

As can be seen from this formulation of LEIBNIZ's law, it is possible to define the concept of identity in terms of the theory of classes.

23. Universal class and null class

As we already know, to any sentential function with one free variable there corresponds the class of all objects satisfying this function. This can now be applied to the following two particular functions:

(I) $$x = x, \qquad x \neq x.$$

The first of these functions is obviously satisfied by every individual (cf. Section 17). The corresponding class,

$$\underset{x}{\mathsf{C}}(x = x),$$

therefore, contains as elements all individuals; we call this class the UNIVERSAL CLASS and denote it by the symbol "V" (or "1"). The second sentential function, on the other hand, is satisfied by no thing. Consequently, the class corresponding to it,

$$\underset{x}{\mathsf{C}}(x \neq x),$$

called the NULL CLASS OR EMPTY CLASS and denoted by "Λ" (or "0"), contains no elements. We may now replace the sentential functions (I) by equivalent functions of the form:

$$x \in K,$$

namely by:

(II) $$x \in \mathsf{V}, \quad x \in \Lambda,$$

the first of which is satisfied by any individual, and the second by none.

Instead of using the general logical concept of individual within a particular mathematical theory, it is sometimes more convenient to specify exactly what is considered an individual thing within the framework of this theory; the class of all those things will then be denoted again by "V" and will be called the UNIVERSE OF DISCOURSE of the theory. In arithmetic, for instance, it is the class of all numbers which forms the universe of discourse.

It should be emphasized that V is the class of all individuals but not the class containing as elements all possible things, thus also classes of first order, second order, and so on. The question arises whether such a class of all possible things exists at all, and more generally, whether we may consider "inhomogeneous" classes not belonging to a particular order and containing as elements individuals as well as classes of various orders. This question is closely related to the most intricate problems of contemporary logic, namely, to the so-called ANTINOMY OF RUSSELL and the THEORY OF LOGICAL TYPES.[2] A discussion of this question would trespass beyond the intended limits of this book. We will only remark here that the need for considering "inhomogeneous" classes occurs hardly ever in the whole of mathematics (except for the general theory of sets), and even more rarely in other sciences.

24. Fundamental relations among classes

Various relations may hold between two classes K and L. It may, for instance, occur that every element of the class K is at the same time an element of the class L in which case the set K is said to be a SUBCLASS OF THE CLASS L or to be INCLUDED IN THE CLASS L, or to HAVE THE RELATION OF INCLUSION TO THE CLASS L; and the class L is said to COMPREHEND THE CLASS K AS A SUBCLASS. This situation is expressed, briefly, by either of the formulas:

$$K \subset L \quad \text{or} \quad L \supset K.$$

By saying that K is subclass of L it is not intended to preclude the possibility of L also being a subclass of K. In other words, K and L may be subclasses of each other and thus have all their elements in common; in this case it follows from a law (given below) of the theory of classes that K and L are identical. If, however, the converse relation does not hold, that is, if every element of the class K is an element of the class L, but if not every element of the class L is an element of the class K, then the class K is said to be a PROPER SUBCLASS OF a PART OF THE CLASS L, and L is said to COMPREHEND K AS A PROPER SUBCLASS OF AS A PART. For example, the set of all integers is a proper subset of the set of all rational numbers; a line comprehends each of its segments as a part.

Two classes K and L are said to OVERLAP or to INTERSECT if they have at least one element in common and if, at the same time,

[2] The concept of logical types introduced by RUSSELL is akin to that of the order of a class, and can even be conceived as a generalization of the latter,—a generalization which refers not only to classes but also to other things, for instance, to relations, which will be considered in the next chapter. The theory of logical types was systematically developed in *Principia Mathematica* (cf. footnote 1 on p. 19).

each contains elements not contained in the other. If two classes have each at least one element (i.e. if they are not empty), but if they have no element in common, they are called MUTUALLY EXCLUSIVE or DISJOINT. A circle, for instance, intersects any straight line drawn through its center, but it is disjoint from any straight line whose distance from the center is greater than the radius. The set of all positive numbers and the set of all rational numbers overlap, but the set of positive and the set of negative numbers are mutually exclusive.

Let us give some examples of laws concerning the relations between classes mentioned above.

For any class K, $K \subset K$.

If $K \subset L$ *and* $L \subset K$, *then* $K = L$.

If $K \subset L$ *and* $L \subset M$, *then* $K \subset M$.

If K is a non-empty subclass of L, and if the classes L and M are disjoint, then the classes K and M are disjoint.

The first of these statements is called the LAW OF REFLEXIVITY for inclusion or the class-theoretical LAW OF IDENTITY. The third is known as the LAW OF TRANSITIVITY for inclusion; together with the fourth statement and others of a similar structure they form a group of statements which are called LAWS OF THE CATEGORICAL SYLLOGISM.

A characteristic property of the universal and null classes in connection with the concept of inclusion is expressed in the following law:

For any class K, $\vee \supset K$ *and* $\wedge \subset K$.

This statement, particularly in view of its second part referring to the null class, seems to many people somewhat paradoxical. In order to demonstrate this second part, let us consider the implication:

if $x \in \wedge$, *then* $x \in K$.

Whatever we substitute here for "x" (and "K"), the antecedent of the implication will be a false sentence, and hence the whole implication a true sentence (the implication—as the mathematicians sometimes say—is satisfied "vacuously"). We may, thus, say that whatever is an element of the class Λ is also an element of the class K, and hence, by the definition of inclusion, that $\Lambda \subset K$. —In an analogous way the first part of the law can be demonstrated.

It is easy to see that between any two classes one of the relations considered here has to hold; the following law is to this effect:

If K and L are two arbitrary classes, then either $K = L$ or K is a proper subclass of L, or K comprehends L as a proper subclass, or K and L overlap, or finally K and L are disjoint; no two of these relations can hold simultaneously.

In order to get a clear intuitive understanding of this law it is best to think of the classes K and L as geometrical figures and to imagine all the possible positions in which these two figures may be with respect to each other.

The relations which have been dealt with in this section may be called the FUNDAMENTAL RELATIONS AMONG CLASSES.[3]

The whole of the old traditional logic (cf. Section 6) can almost entirely be reduced to the theory of the fundamental relations among classes, that is, to a small fragment of the entire theory of classes. Outwardly these two disciplines differ by the fact that, in the old logic, the concept of a class does not appear explicitly. Instead of saying, for instance, that the class of horses is contained in the class of mammals, one used to say in the old logic that the property of being a mammal belongs to all horses, or, simply, that every horse is a mammal. The most important laws of traditional logic are those of the categorical syllogism which correspond precisely to the laws of the theory of classes that we stated above and named after them. For example, the first of the laws of syllogism given above assumes the following form in the old logic:

If every M is P and every S is M, then every S is P.

[3] These relations were first investigated in an exhaustive manner by the French mathematician J. D. GERGONNE (1771–1859).

This is the most famous of the laws of traditional logic, known as the law of the syllogism BARBARA.

25. Operations on classes

We shall now concern ourselves with certain operations which, if performed on given classes, yield new classes.

Given any two classes K and L, one can form a new class M which contains as its elements those, and only those, things which belong to at least one of the classes K and L; the class M, one might say, results from the class K by adjoining to it the elements of the class L. This operation is called ADDITION OF CLASSES, and the class M is referred to as the SUM OR UNION OF THE CLASSES K AND L, designated by the symbol:

$$K \cup L \quad (\text{or} \quad K + L).$$

Another operation on two classes K and L, called MULTIPLICATION OF CLASSES, consists in forming a new class M whose elements are those, and only those, things which belong to both K and L; this class M is called the PRODUCT OR INTERSECTION OF THE CLASSES K AND L and is designated by the symbol:

$$K \cap L \quad (\text{or} \quad K \cdot L).$$

These two operations are frequently applied in geometry; sometimes it is very convenient to define with their help new kinds of geometrical figures. Suppose, for instance, we know already what is meant by a pair of supplementary angles; then the half-plane—that is, the straight angle—may be defined as the union of two supplementary angles (an angle here being considered as an angular region, that is, as a part of the plane, bounded by the two half-lines which are called the legs of the angle). Or, if we take an arbitrary circle and an angle whose vertex lies in the center of the circle, then the intersection of these two figures is a figure called a circular sector.

Let us add two more examples from the field of arithmetic: the sum of the set of all positive numbers and of the set of all negative numbers is the set of all numbers different from 0; the intersection of the set of all even numbers and of the set of all prime numbers is the set having as its sole element the number 2, this number being the only even prime number.

The addition and multiplication of classes are governed by various laws. Some of these are completely analogous to the corresponding theorems of arithmetic concerning the addition and multiplication of numbers—and it is for this very reason that the terms "addition" and "multiplication" have been chosen for the above operations; as an example we mention the COMMUTATIVE and ASSOCIATIVE LAWS of addition and multiplication of classes:

For any classes K and L, $K \cup L = L \cup K$ *and* $K \cap L = L \cap K$.

For any classes K, L and M, $K \cup (L \cup M) = (K \cup L) \cup M$ *and* $K \cap (L \cap M) = (K \cap L) \cap M$.

The analogy with the corresponding arithmetical theorems becomes evident when we replace the symbols "$\cup$" and "$\cap$" by the usual signs of addition and multiplication, "+" and "·".

Other laws, however, deviate considerably from those of arithmetic; the LAW OF TAUTOLOGY constitutes a characteristic example:

For any class K, $K \cup K = K$ *and* $K \cap K = K$.

This law becomes obvious on reflecting upon the meaning of the symbols "$K \cup K$" and "$K \cap K$"; if, for instance, one adds to the elements of the class K the elements of the same class, one does not really add anything, and the resulting class is again the same class K.

We want to mention one other operation, which differs from those of addition and multiplication inasmuch as it can be performed, not on two classes, but only on one class. It is the operation which consists in forming, from a given class K, the so-called COMPLEMENT OF THE CLASS K, that is, the class of all things not belonging to the class K; the complement of the class K is denoted by:

$$K'.$$

If K, for instance, is the set of all integers, all fractions and irrational numbers belong to the set K'.

As examples of laws which concern the concept of complement

and establish its connection with concepts considered earlier, we give the following two statements:

$$\text{For every class } K, \quad K \cup K' = \vee.$$
$$\text{For every class } K, \quad K \cap K' = \wedge.$$

The first of these is called the class-theoretical LAW OF EXCLUDED MIDDLE, and the second the class-theoretical LAW OF CONTRADICTION.

The relations between classes and the operations on classes with which we have just become acquainted, and also the concepts of the universal class and the null class, are treated in a special part of the theory of classes; since the laws concerning those relations and operations tend to have the character of simple formulas reminiscent of those of arithmetic, this part of the theory is known as the CALCULUS OF CLASSES.

26. Equinumerous classes, cardinal number of a class, finite and infinite classes; arithmetic as a part of logic

*Among the remaining concepts which form the subject of investigation of the theory of classes there is one group which deserves particular attention and which comprises such concepts as equinumerous classes, cardinal number of a class, finite and infinite classes. They are, unfortunately, rather involved concepts which can only be superficially discussed here.

As an example of two EQUINUMEROUS OR EQUIVALENT CLASSES, we may consider the sets of the fingers of the right and of the left hands; these sets are equinumerous, because it is possible to pair off the fingers of both hands in such a manner that (i) every finger occurs in just one pair, and (ii) every pair contains just one finger of the left hand and just one finger of the right hand. In a similar sense, the following three sets, for instance, are equinumerous: the set of all vertices, the set of all sides, and the set of all angles of a polygon. Later, in Section 33, we shall be able to give an exact and general definition of this concept of equinumerous classes.

Now let us consider an arbitrary class K; there exists, no doubt, a property belonging to all classes equinumerous to K and to no other classes (namely, the property of being equinumerous with K); this property is called the CARDINAL NUMBER, or the NUMBER OF ELEMENTS, or the POWER OF THE CLASS K. This can also be expressed more briefly and precisely, though perhaps in an even more abstract manner: The cardinal number of a class K is the class of all classes equinumerous with K. It follows from this that two classes K and L have the same cardinal number if, and only if, they are equinumerous.

With regard to the number of their elements, classes are classified into finite and infinite ones. Among the former, we distinguish between classes consisting of exactly one element, of two, of three elements, and so on. These terms are most easily definable on the basis of arithmetic. Indeed, let n be an arbitrary natural number (that is, a non-negative integer); then we shall say that THE CLASS K CONSISTS OF n ELEMENTS, if K is equinumerous with the class of all natural numbers less than n. In particular, a class consists of 2 elements, if it is equinumerous with the class of all natural numbers less than 2, i.e., to the class consisting of the numbers 0 and 1. Similarly, a class consists of 3 elements if it is equinumerous with the class containing the numbers 0, 1 and 2 as elements. In general, we shall call a class K FINITE if there exists a natural number n such that the class K consists of n elements, otherwise INFINITE.

It has, however, been recognized that there is still another possible procedure. All the terms which have just been considered can be defined in purely logical terms, without resorting at all to any expressions belonging to the field of arithmetic. We may, for instance, say that the class K consists of exactly one element, if this class satisfies the following two conditions: (i) there is an x such that $x \in K$; (ii) for any y and z, if $y \in K$ and $z \in K$, then $y = z$ (these two conditions may also be replaced by a single one: "there is exactly one x such that $x \in K$"; cf. Section 20). Analogously, we can define the phrases: "the class K consists of two elements", "the class K consists of three elements", and so on. The problem becomes much more difficult when we turn to the question of defining the terms "finite class" and "infinite class"; but also in these cases the efforts of solving the problem positively have been successful (cf. Section 33), and thereby all the concepts under consideration have been included within the range of logic.

This circumstance has a most interesting consequence of far-reaching importance; for it turns out that the notion of number itself and likewise all other arithmetical concepts are definable within the field of logic. It is, indeed, easy to establish the meaning of symbols designating individual natural numbers, such as "0", "1", "2", and so on. The number 1, for instance, can be defined as the number of elements of a class which consist of exactly one element. (A definition of this kind seems to be incorrect and contains apparently a vicious circle, since the word "one", which is about to be defined, occurs in the definiens; but actually no error is committed because the phrase "the class consists of exactly one element" is considered as a whole and its meaning has been defined previously.) Nor is it hard to define the general concept of a natural number: a natural number is the cardinal number of a finite class. We are, further, in a position to define all operations on natural numbers, and to extend the concept of number by the introduction of fractions, negative and irrational numbers, without, at any place, having to go beyond the limits of logic. Furthermore, it is possible to prove all the theorems of arithmetic on the basis of laws of logic alone (with the qualification that the system of logical laws must first be enriched by the inclusion of a statement which is intuitively less evident than the others, namely, the so-called AXIOM OF INFINITY, which states that there are infinitely many different things). This entire construction is very abstract, it cannot easily be popularized and does not fit into the framework of an elementary presentation of arithmetic; in this book we also do not attempt to adapt ourselves to this conception and treat numbers as individuals and not as properties or classes of classes. But the mere fact that it has been possible to develop the whole of arithmetic, including the disciplines erected upon it—algebra, analysis, and so on—, as a part of pure logic, constitutes one of the grandest achievements of recent logical investigations.[4]*

[4] The fundamental ideas in this field are due to FREGE (cf. footnote 2 on p. 19); he developed them for the first time in his interesting book: *Die Grundlagen der Arithmetik* (Breslau 1884). FREGE's ideas found their systematic and exhaustive realization in WHITEHEAD and RUSSELL's *Principia Mathematica* (cf. footnote 1 on p. 19).

Exercises

1. Let K be the set of all numbers less than $\frac{3}{4}$; which of the following formulas are true:

$$0 \in K, \qquad 1 \in K, \qquad \tfrac{2}{3} \in K, \qquad \tfrac{3}{4} \in K, \qquad \tfrac{4}{5} \in K \quad ?$$

2. Consider the following four sets:

(a) the set of all positive numbers,

(b) the set of all numbers less than 3,

(c) the set of all numbers x such that $x + 5 < 8$,

(d) the set of all numbers x satisfying the sentential function "$x < 2x$".

Which of these sets are identical, and which are distinct?

3. What name is given in geometry to the set of all points in space whose distance from a given point (or from a given straight line) does not exceed the length of a given line segment?

4. Let K and L be two concentric circles, the radius of the first being smaller than that of the second. Which of the relations discussed in Section 20 holds between these circles? Does the same relation hold between the circumferences of the circles?

5. Draw two squares K and L so that they stand in one of the following relations:

(a) $K = L$,

(b) the square K is a part of the square L,

(c) the square K comprehends the square L as a part,

(d) the squares K and L overlap,

(e) the squares K and L are disjoint.

Which of these cases are eliminated, (i) if the squares are congruent, or (ii) if not the squares but only their perimeters are considered?

6. Let x and y be two arbitrary numbers, with $x < y$. It is well known that the set of numbers which are not smaller than

x and not larger than y is called the interval with the endpoints x and y; it is denoted by the symbol "$[x, y]$".

Which of the formulas below are correct:

(a) $[3, 5] \subset [3, 6]$,

(b) $[4, 7] \subset [5, 10]$,

(c) $[-2, 4] \supset [-3, 5]$,

(d) $[-7, 1] \supset [-5, -2]$?

Which of the fundamental relations hold between the intervals:

(e) $[2, 4]$ and $[5, 8]$,

(f) $[3, 6]$ and $[3\frac{1}{2}, 5\frac{1}{2}]$,

(g) $[1\frac{1}{2}, 7]$ and $[-2, 3\frac{1}{2}]$?

7. Is the following sentence (which has the same structure as the laws of syllogism given in Section 24) true:

if K is disjoint from L and L disjoint from M, then K is disjoint from M ?

8. Translate the following formulas into terms of ordinary language:

(a) $(x = y) \leftrightarrow \bigwedge_K [(x \in K) \leftrightarrow (y \in K)]$,

(b) $(K = L) \leftrightarrow \bigwedge_x [(x \in K) \leftrightarrow (x \in L)]$.

What laws mentioned in Section 22 and 24 find their expression in these formulas? What alterations on both sides of the equivalence (b) would be required in order to arrive at a definition of the symbol "$\subset$" or "$\supset$"?

9. Let ABC be an arbitrary triangle, with an arbitrary point D lying on the segment BC. What figures are formed by the sum of the two triangles ABD and ACD and by their product? Express the answer in formulas.

10. Represent an arbitrary square:

(a) as the sum of two trapezoids,

(b) as the intersection of two triangles.

11. Which of the formulas below are true (compare exercise 6):

(a) $[2, 3\frac{1}{2}] \cup [3, 5] = [2, 5]$,

(b) $[-1, 2] \cup [0, 3] = [0, 2]$,

(c) $[-2, 8] \cap [3, 7] = [-2, 8]$,

(d) $[2, 4\frac{1}{2}] \cap [3, 5] = [2, 3]$?

In those formulas which are false correct the expression on the right of the symbol "$=$".

12. Let K and L be two arbitrary classes. What classes are $K \cup L$ and $K \cap L$ in case $K \subset L$? In particular, what classes are $K \cup \mathrm{V}, K \cap \mathrm{V}, \Lambda \cup L$ and $\Lambda \cap L$?

Hint: In answering the second question keep in mind a law of Section 24 concerning the classes V and Λ.

13. Try to show that any classes K, L and M satisfy the following formulas:

(a) $K \subset K \cup L$ and $K \supset K \cap L$,

(b) $K \cap (L \cup M) = (K \cap L) \cup (K \cap M)$
and $K \cup (L \cap M) = (K \cup L) \cap (K \cup M)$,

(c) $(K')' = K$,

(d) $(K \cup L)' = K' \cap L'$ and $(K \cap L)' = K' \cup L'$.

The formulas (a) are called the LAWS OF SIMPLIFICATION (for addition and multiplication of classes); the formulas (b) are the DISTRIBUTIVE LAWS (for the multiplication of classes with respect to addition and for addition with respect to multiplication); the formula (c) is the LAW OF DOUBLE COMPLEMENT; and, finally, the formulas (d) are the class-theoretical LAWS OF DE MORGAN.[5] Which of these laws correspond to theorems of arithmetic?

Hint: In order to prove the first of the formulas (d), for instance, it is sufficient to show that the classes $(K \cup L)'$ and $K' \cap L'$ consist entirely of the same elements (cf. Section 24). For this purpose,

[5] Cf. footnote 6 on p. 52.

we must, using the definitions of Section 25, make clear to ourselves when a thing x belongs to the class $(K \cup L)'$ and when it belongs to the class $K' \cap L'$.

*14. Between the laws of sentential calculus given in Sections 12 and 13 and in Exercise 14 of Chapter II, on the one hand, and the laws of the calculus of classes given in Sections 24 and 25 and in the preceding exercise, on the other, there subsists a far-reaching similarity in structure (which is indicated in the analogy in their names). Describe in detail wherein this similarity lies, and try to find a general explanation of this phenomenon.

In Section 14, we became acquainted with the law of contraposition of sentential calculus; formulate the analogous law of the calculus of classes.

15. With the help of the symbol:

$$\underset{x}{\mathsf{C}}$$

introduced in Section 22 we can write the definition of the sum of two classes in the following way:

$$K \cup L = \underset{x}{\mathsf{C}}[(x \in K) \vee (x \in L)];$$

but it is also possible to restate this definition in the usual form of an equivalence (without the use of that symbol):

$$[x \in (K \cup L)] \leftrightarrow [(x \in K) \vee (x \in L)].$$

Formulate analogously in two ways the definitions of the universal class, of the null class, of the product of two classes, and of the complement of a class.

*16. Is there a polygon, in which the set of all sides is equinumerous with the set of all diagonals?

*17. Lay down definitions of the following expressions, using terms from the field of logic exclusively:

(a) *the class K consists of two elements,*

(b) *the class K consists of three elements.*

*18. Consider the following three sets:

(a) the set of all natural numbers greater than 0 and less than 4,

(b) the set of all rational numbers greater than 0 and less than 4,

(c) the set of all irrational numbers greater than 0 and less than 4.

Which of these sets are finite and which are infinite?
Give further examples of finite and infinite sets of numbers.

· V ·

ON THE THEORY OF RELATIONS

27. Relations, their domains and counter-domains; relations and sentential functions with two free variables

In the previous chapters we have already met with a few RELATIONS between things. As examples of relations between two things we may take, for instance, identity (equality) and diversity (inequality). We sometimes read the formula:

$$x = y$$

as follows:

x has the relation of identity to y

or also:

the relation of identity holds between x and y,

and we say that the symbol "=" designates the relation of identity. In an analogous way, the formula:

$$x \neq y$$

is sometimes read:

x has the relation of diversity to y

or:

the relation of diversity holds between x and y,

and one says that the symbol "≠" designates the relation of diversity. We have further encountered certain relations holding between classes, namely, the relations of inclusion, of overlapping, of disjointness, and so on. We will now discuss several concepts belonging to the general THEORY OF RELATIONS, which constitutes a special and very important part of logic, and in which relations of an entirely arbitrary character are considered and general laws concerning them are established.[1]

To facilitate our considerations, we introduce special variables "R", "S", $\cdots$ which serve to denote relations. In place of such phrases as:

the thing x has the relation R to the thing y

and:

the thing x does not have the relation R to the thing y

we shall employ symbolic abbreviations:

$$x \, R \, y$$

and (to use the negation sign of sentential calculus, cf. Section 13)

$$\sim(x \, R \, y),$$

respectively.

Any thing having the relation R to some thing y we call a PREDECESSOR WITH RESPECT TO THE RELATION R; any thing y for which there is a thing x such that

$$x \, R \, y$$

is called a SUCCESSOR WITH RESPECT TO THE RELATION R. The class of all predecessors with respect to the relation R is known as the DOMAIN and the class of all successors as the COUNTER-DOMAIN (or CONVERSE DOMAIN) OF THE RELATION R. Thus, for example, any individual is both a predecessor and a successor with respect to the relation of identity, so that the domain and counter-domain of this relation are both the universal class.

In the theory of relations—just as in the theory of classes—we may distinguish relations of different orders. The RELATIONS

[1] DE MORGAN and PEIRCE (cf. footnotes 6 on p. 52 and 2 on p. 14) were first to develop the theory of relations, especially that part of it known as the calculus of relations (cf. Section 28). Their work was systematically expanded and completed by the German logician E. SCHRÖDER (1841–1902). SCHRÖDER's *Algebra und Logik der Relative* (Leipzig 1895), which appeared as the third volume of his comprehensive work *Vorlesungen über die Algebra der Logik*, is still the only exhaustive account of the calculus of relations.

OF THE FIRST ORDER are those which hold between individuals; the RELATIONS OF THE SECOND ORDER are those which hold between classes, or relations, of the first order; and so on. The situation is here all the more complicated as we must often consider "mixed" relations whose precedessors are, say, individuals, and its successors classes, or whose predecessors are, for instance, classes of the first order and its successors classes of the second order. The most important example of a relation of this kind is the relation which holds between an element and a class to which it belongs; as we recall from Section 21, this relation is denoted by the symbol "ϵ".—As in the case of classes, our considerations concerning relations will refer primarily to those of the first order, although the concepts discussed here can and, in a few cases, will be applied to relations of higher orders.

We assume that, to every sentential function with two free variables "x" and "y", there corresponds a relation holding between the things x and y if, and only if, they satisfy the given sentential function; in this connection it is said of a sentential function with the free variables "x" and "y" that it expresses a relation between the things x and y. Thus, for instance, the sentential function:

$$x + y = 0$$

expresses the relation of having the opposite sign or, briefly, of being opposite; the numbers x and y have the relation of being opposite if, and only if, $x + y = 0$. If we denote this relation by the symbol "O", then the formulas:

$$x \,\mathrm{O}\, y$$

and

$$x + y = 0$$

are equivalent. Similarly, any sentential function containing the symbols "x" and "y" as the only free variables may be transformed into an equivalent formula of the form:

$$x \, R \, y$$

where, in place of "R", we have a constant which designates some relation. The formula:

$$x\ R\ y$$

may, therefore, be considered as the general form of a sentential function with two free variables, just as the formula:

$$x \in K$$

could be looked upon as the general form of a sentential function with one free variable (cf. Section 22).

28. Calculus of relations

The theory of relations is one of the farthest developed branches of mathematical logic. One part of it, the CALCULUS OF RELATIONS, is akin to the calculus of classes, its principal object being the establishment of formal laws governing the operations by means of which other relations are constructed from given ones.

In the calculus of relations we consider, in the first place, a group of concepts which are exact analogues of those of the calculus of classes; they are usually denoted by the same symbols and governed by quite similar laws. (In order to avoid ambiguity, we might, of course, employ a different set of symbols in the calculus of relations, taking, for instance, the symbols of the calculus of classes and placing a dot over each.)

We have thus in the calculus of relations two special relations, the UNIVERSAL RELATION $\vee$ and the NULL RELATION $\wedge$, the first of which holds between any two individuals, and the second between none.

We have, further, various relations between relations, for instance, the RELATION OF INCLUSION; we say that the relation R is INCLUDED in the relation S, in symbols:

$$R \subset S,$$

if, whenever R holds between two things, S holds between them likewise; or, in other words, if, for any x and y, the formula:

$$x\ R\ y$$

implies:

$$x\ S\ y.$$

We know, for instance, from arithmetic that, whenever

$$x < y,$$

then

$$x \neq y;$$

hence the relation of being smaller is included in the relation of diversity.

If, at the same time,

$$R \subset S \quad \text{and} \quad S \subset R,$$

that is to say, if the relations R and S hold between the same things, then they are identical:

$$R = S.$$

We have, further, the SUM OR UNION OF TWO RELATIONS R AND S, in symbols:

$$R \cup S,$$

and the PRODUCT OR INTERSECTION OF R AND S, in symbols:

$$R \cap S.$$

The first, $R \cup S$, holds between two things if, and only if, at least one of the relations R and S holds between them; in other words, the formula:

$$x(R \cup S)y$$

is equivalent to the condition:

$$x\,R\,y \quad or \quad x\,S\,y.$$

Similarly the product of two relations is defined, using only the word "*and*" instead of "*or*". Thus, for example, if R is the relation of fatherhood (that is, a relation holding between two persons x and y if, and only, x is the father of y), and S the relation of motherhood, then $R \cup S$ is the relation of parenthood, while $R \cap S$ is, in this case, the null relation.

We have, finally, the NEGATION or the COMPLEMENT OF A RELATION R denoted by:

$$R'.$$

It is a relation which holds between two things if, and only if, the relation R does not hold between them; in other words, for any x and y, the formulas:

$$x\,R'\,y \quad \text{and} \quad \sim(x\,R\,y)$$

are equivalent. It should be noted that, if a relation is designated by a constant, then its complement is frequently denoted by the symbol obtained from that constant by crossing it by a vertical or oblique bar. The negation of the relation $<$, for example, is usually denoted by "$\nless$", and not by "$<'$".

In the calculus of relations there occur also entirely new concepts, without analogues in the calculus of classes.

We have here, first, two special relations, IDENTITY and DIVERSITY between individuals (which are, incidentally, familiar to us from earlier considerations). In the calculus of relations they are denoted by special symbols, e.g., "I" and "D", and not by the symbols "$=$" and "$\neq$" used in other parts of logic. We write, thus:

$$x\,\mathsf{I}\,y \quad \text{and} \quad x\,\mathsf{D}\,y$$

instead of:

$$x = y \quad \text{and} \quad x \neq y.$$

The symbols "$=$" and "$\neq$" are used in the calculus of relations only to denote the identity and diversity between relations.

We have here, further, a very interesting and important new operation, with the help of which we form, from two relations R and S, a third relation called the RELATIVE PRODUCT OF COMPOSITION OF R AND S (as opposed to it, the ordinary product is sometimes called the ABSOLUTE PRODUCT). The relative product of R and S is denoted by the symbol:

$$R/S;$$

it holds between two things x and y if, and only if, there exists a thing z such that we have at the same time:

$$x\,R\,z \quad \text{and} \quad z\,S\,y.$$

Thus, for instance, if R is the relation of being husband and S the relation of being daughter, then R/S holds between two persons

x and y if there is a person z such that x is husband of z and z is daughter of y; the relation R/S, therefore, coincides with the relation of being son-in-law.—We have here, in addition, another operation of a similar character, whose result is called the RELATIVE SUM OF TWO RELATIONS. This operation does not play a very great role and will not be defined here.

Finally, we have an operation similar to that of forming R', namely, an operation with the help of which, from a relation R, we form a new relation called the CONVERSE OF R and denoted by:

$$\breve{R}.$$

The relation $\breve{R}$ holds between x and y if, and only if, R holds between y and x. If a relation is denoted by a constant, then for denoting its converse we often employ the same symbol printed in the opposite direction. The converse of the relation $<$, for instance, is the relation $>$, since, for any x and y, the formulas:

$$x < y \quad \text{and} \quad y > x$$

are equivalent.

In view of the rather specialized character of the calculus of relations, we shall here not go any further into the details of it.

29. Some properties of relations

We now turn to that part of the theory of relations whose task it is to single out and investigate special kinds of relations with which one meets frequently in other sciences and, in particular, in mathematics.

We shall call a relation R REFLEXIVE IN THE CLASS K, if every element x of the class K has the relation R to itself:

$$x\, R\, x;$$

if, on the other hand, no element of this class has the relation R to itself:

$$\sim(x\, R\, x),$$

then the relation R is said to be IRREFLEXIVE IN THE CLASS K. The relation R is called SYMMETRICAL IN THE CLASS K if, for any two elements x and y of the class K, the formula:

$$x\, R\, y$$

always implies the formula:

$$y\ R\ x.$$

If, however, the formula:

$$x\ R\ y$$

always implies:

$$\sim(y\ R\ x),$$

then the relation R is said to be ASYMMETRICAL IN THE CLASS K. The relation R is called TRANSITIVE IN THE CLASS K if, for any three elements x, y and z of the class K the conditions:

$$x\ R\ y \quad \text{and} \quad y\ R\ z$$

always imply:

$$x\ R\ z.$$

If, finally, for any two different elements x and y of the class K, at least one of the formulas:

$$x\ R\ y \quad \text{and} \quad y\ R\ x$$

holds, that is, if the relation R subsists between two arbitrary distinct elements of K in at least one direction, the relation is called CONNECTED IN THE CLASS K.

In case K is the universal class (or, at any rate, the universe of discourse of the science in which we happen to be interested—cf. Section 23) we usually speak, more briefly, not of relations reflexive, symmetrical, and so on, in the class K, but simply of reflexive relations, symmetrical relations, and so on.

30. Relations which are reflexive, symmetrical and transitive

The above-mentioned properties of relations frequently occur in groups. Very common, for instance, are those relations which are reflexive, symmetrical, and transitive as well. A typical example of this type is the relation of identity; Law II of Section 17 expresses that this relation is reflexive, by Law III identity is a symmetrical relation, and according to Law IV it is transitive (and this accounts for the names of these laws that were given in Section 17). Numerous other examples of relations of this kind may be found within the field of geometry. Congruence, for instance, is a reflexive relation in the set of all line segments (or of arbitrary geometrical configurations), since every segment is congruent to itself; it is symmetrical, since, if a segment is congruent to another segment, the other is congruent to the first; and, finally, it is transitive, since, if the segment A is congruent to the segment B, and B to C, then the segment A is also congruent to the segment C. The same three properties belong to the relations of similarity among polygons or of parallelism among straight lines (assuming any line to be parallel to itself), or—outside the domain of geometry—to the relations of being equally old among people, or of synonymity among words.

Every relation which is at the same time reflexive, symmetrical and transitive is thought of as some kind of equality. Instead of saying, therefore, that such a relation holds between two things, one can, in this sense, also say that these things are equal in such and such a respect, or—in a more precise mode of speech—that certain properties of these things are identical. Thus, instead of stating that two segments are congruent, or two people equally old, or two words synonymous, it may just as well be stated that the segments are equal in respect of their length, that the people have the same age, or that the meanings of the words are identical.

*By way of an example we will give an indication of how it is possible to establish a logical basis for such a mode of expression. For this purpose let us consider the relation of similarity among polygons. We will denote the set of all polygons similar to the given polygon P (or, to use a slightly more current terminology, the common property which belongs to all polygons similar to P and to no others) as the shape of the polygon P. Thus shapes are certain sets of polygons (or properties of polygons; cf. remarks at the end of Section 22). Making use of the fact mentioned above that the relation of similarity is reflexive, symmetrical and transitive, we can now easily show that every polygon belongs to one and only one such set, that two similar polygons belong always to the same set, and that two polygons which are not similar belong to different sets. From this it follows at once that the two statements:

the polygons P and Q are similar

and

the polygons P and Q have the same shape (that is, *the shapes of P and Q are identical*)

are equivalent.

The reader will notice immediately that, in the course of the preceding considerations, we have once before applied an analogous procedure, namely in Section 26 in making the transition from the expression:

the classes K and L are equinumerous

to the equivalent one:

the classes K and L have the same cardinal number.

It can be shown with little difficulty that the same procedure is applicable to any reflexive, symmetrical and transitive relation. There is even a logical law, called the PRINCIPLE OF ABSTRACTION, that supplies a general theoretical foundation for the procedure which we have been considering, but we shall here forego the exact formulation of this principle.*

There is, so far, no universally accepted term denoting the totality of relations which are at the same time reflexive, symmetrical and transitive. Sometimes they have generally been called EQUALITIES or EQUIVALENCES. But the term "equality" is also sometimes reserved for particular relations of the category under consideration, and two things are then called equal if such a relation holds between them. For instance, in geometry, as has been pointed out in Section 19, congruent segments are often referred to as equal segments. We will emphasize here once more that it is preferable to avoid such expressions altogether; their use merely leads to ambiguities, and it violates the convention in accordance with which we consider the terms "equality" and "identity" as synonymous.

31. Ordering relations; examples of other relations

Another very common kind of relation is represented by those which are asymmetrical, transitive and connected in a given class *K* (they must then, as can be shown, also be irreflexive in the class

K). Of a relation with these properties we say that it **ESTABLISHES AN ORDER IN THE CLASS** K; we say also that the class K is **ORDERED BY THE RELATION** R. Consider, for example, the relation of being smaller (or, the relation *less than*, as we shall say occasionally); it is asymmetrical in any set of numbers, for, if x and y are any two numbers and if

$$x < y$$

then

$$y \nless x, \quad \text{i.e.} \quad \sim(y < x);$$

it is transitive, since the formulas:

$$x < y \quad \text{and} \quad y < z$$

always imply:

$$x < z;$$

finally, it is connected, since, of any two distinct numbers, one must be smaller than the other (and it is also irreflexive, since no number is smaller than itself). Any set of numbers, therefore, is ordered by the relation of being smaller. Likewise, the relation of being greater represents another ordering relation for any set of numbers.

Let us now consider the relation of being older. One can easily verify that this relation is irreflexive, asymmetrical and transitive in any given set of people. However, it is not necessarily connected; for it can happen, perchance, that the set contains two people having exactly the same age, that is to say, who were born at the same moment, so that the relation of being older does not hold between them in either direction. If, on the other hand, we consider a set of people in which no two are of exactly the same age, the relation of being older establishes an order in that set.

Many instances of relations are known that belong to neither of the two categories discussed in the present section and in the preceding one. Let us consider a few examples.

The relation of diversity is irreflexive in any set of things, since no thing is different from itself; it is symmetrical, for, if

$$x \neq y,$$

then we also have

$$y \neq x;$$

it fails to be transitive, however, since the formulas:

$$x \neq y \quad \text{and} \quad y \neq z$$

do not imply the formula:

$$x \neq z;$$

it is, on the other hand, connected, as can be seen at once.

The relation of inclusion between classes, by the law of identity and one of the laws of syllogism (cf. Section 24), is reflexive and transitive; it is, further, neither symmetrical nor asymmetrical, since the formula:

$$K \subset L$$

neither implies nor excludes the formula:

$$L \subset K$$

(these two formulas are fulfilled simultaneously if, and only if, the classes K and L are identical); finally, it can be seen with ease that it is not connected. Thus, the relation of inclusion differs in its properties from other relations thus far considered.

32. One-many relations or functions

We will now deal in some detail with another particularly important category of relations. A relation R is called a ONE-MANY OR FUNCTIONAL RELATION or simply a FUNCTION if, to every thing y, there corresponds at most one thing x such that $x\ R\ y$; in other words, if the formulas:

$$x\ R\ y \quad \text{and} \quad z\ R\ y$$

always imply the formula:

$$x = z.$$

The successors with respect to the relation R, that is, those things y for which there actually are things x such that

$$x\ R\ y,$$

are the ARGUMENT VALUES, the predecessors are the FUNCTION VALUES or, simply, the VALUES OF THE FUNCTION R. Let R be an arbitrary function, y any one of its argument values; the unique value x of the function corresponding to the value y of the argument we will denote by the symbol "$R(y)$"; consequently we replace the formula:

$$x\,R\,y$$

by:

$$x = R(y).$$

It has become the custom, especially in mathematics, to use, not the variables "R", "S", $\cdots$, but other letters such as "f", "g", $\cdots$ to denote functional relations, so that we find formulas like these:

$$x = f(y), \qquad x = g(y), \cdots;$$

the formula:

$$x = f(y),$$

for instance, is read as follows:

the function f assigns (or *correlates*) *the value x to the argument value y*

or

x is that value of the function f which corresponds to (or *is correlated with*) *the argument value y.*

(There is also another custom, of using the variable "x" for denoting the argument value and the variable "y" for denoting the value of the function. We shall not adhere to this custom, and continue to use "x" and "y" in the opposite order, because this is more convenient in connection with the general notation used in the theory of relations.)

In many elementary textbooks of algebra a definition of the concept of a function is to be found that is quite different from the definition adopted here. The functional relation is there characterized as a relation between two "variable" quantities or numbers: the "independent variable" and the "dependent variable", which depend upon each other in so far as a change of the first effects a change of the second. Definitions of this kind should no longer be employed today, since they are incapable of standing up to any logical criticism; they are the remains of a period in which one tried to distinguish between "constant" and "variable" quantities (cf. Section 1). He who desires to comply with the requirements of contemporary science and yet does not wish to break away completely from tradition, may, however, retain the old terminology and use, beside the terms "argument value" and "function value", the expressions "value of the independent variable" and "value of the dependent variable".

The simplest example of a functional relation is represented by the ordinary relation of identity. As an example of a function from everyday life let us take the relation expressed by the sentential function:

x is father of y.

It is a functional relation, since, to every person y, there exists but one person x who is father of y. In order to indicate the functional character of this relation, we insert the word *"the"* in the above formulation:

x is the father of y,

instead of which we might also write:

x is identical with the father of y.

Such an alteration of the original expression, involving the insertion of the definite article, serves, in ordinary language, exactly the same purpose as the transition from the formula:

$$x \, R \, y$$

to the formula:

$$x = R(y)$$

in our symbolism.

The concept of a function plays a most important role in the mathematical sciences. There are whole branches of higher mathematics devoted exclusively to the study of certain kinds of functional relations. But also in elementary mathematics, especially in algebra and trigonometry, we find an abundance of functional relations. Examples are the relations expressed by such formulas as:

$$x + y = 5,$$

$$x = y^2,$$

$$x = \log_{10} y,$$

$$x = \sin y,$$

and many others. Let us consider the second of these formulas more closely. To every number y, there corresponds only one number x such that $x = y^2$, so that the formula really does represent a functional relation. Argument values of this function are arbitrary numbers, values of the function, however, only non-negative numbers. If we denote this function by the symbol "f", the formula:

$$x = y^2$$

assumes the form:

$$x = f(y).$$

Evidently "x" and "y" may here be replaced by symbols designating definite numbers. Since, for instance,

$$4 = (-2)^2,$$

it may be asserted that

$$4 = f(-2);$$

thus, 4 is the value of the function f corresponding to the argument value -2.

On the other hand, and again in elementary mathematics already, we encounter numerous relations which are not functions. For example, the relation of being smaller is certainly not a function, since, to every number y, there are infinitely many numbers x such that

$$x < y.$$

Nor is the relation between the numbers x and y expressed by the formula:

$$x^2 + y^2 = 25$$

a functional relation, since, to one and the same number y, there may correspond two different numbers x for which the formula is valid; corresponding to the number 4, for instance, we have both the numbers 3 and -3. It may be noted that relations between numbers which, like the one just considered, are expressed by equations and correlate with one number y two or more numbers x are sometimes called in mathematics two- or many-valued functions (in opposition to single-valued functions, that is, to functions in the ordinary meaning). It seems, however, inexpedient—at least on an elementary level—to denote such relations as functions, for this only tends to blot out the essential difference between the notion of a function and the more general one of a relation.

Functions are of particular significance as far as the application of mathematics to the empirical sciences is concerned. Whenever we inquire into the dependence between two kinds of quantities occurring in the external world, we strive to give this dependence the form of a mathematical formula, which would permit us to determine exactly the quantity of the one kind by the corresponding quantity of the other; such a formula always represents some functional relation between the quantities of two kinds. As an example let us mention the well-known formula from physics:

$$s = 16.1\, t^2$$

expressing the dependence of the distance s, covered by a freely falling body, upon the time t of its fall (the distance being measured in feet and the time in seconds).

*In conclusion of our remarks on functional relations we want to emphasize that the concept of a function which we are considering now differs essentially from the concepts of a sentential and of a designatory function known from Section 2. Strictly speaking, the terms "sentential function" and "designatory function" do not belong to the domain of logic or mathematics; they denote certain categories of expressions which serve to compose logical and mathematical statements, but they do not denote things treated of in those statements (cf. Section 9). The term "function" in its new sense, on the other hand, is an expression of a purely logical character; it designates a certain type of things dealt with in logic and mathematics. There is, no doubt, a connection between these concepts, which may be described roughly as follows. If the variable "x" is joined by the symbol "$=$" to a designatory function containing "y" as the only variable, e.g. to "$y^2 + 2y + 3$", then the resulting formula (which is a sentential function):

$$x = y^2 + 2y + 3$$

expresses a functional relation; or, in other words, the relation holding between those and only those numbers x and y which satisfy this formula is a function in the new sense. This is one of the reasons why these concepts are so often confused.*

33. One-one relations or biunique functions, and one-to-one correspondences

Among the functional relations particular attention should be paid to the so-called ONE-ONE RELATIONS OR BIUNIQUE FUNCTIONS, that is, to those functional relations in which not only to every argument value y only one function value x is correlated, but also conversely only one argument value y corresponds to every value x of the function; they might also be defined as those relations which have the property that their converses (cf. Section 28) as well as the relations themselves are one-many.

If f is a biunique function, K an arbitrary class of its argument values, and L the class of function values correlated with the elements of K, we say that the function f MAPS THE CLASS K ON THE CLASS L IN A ONE-TO-ONE MANNER, or that it ESTABLISHES A ONE-TO-ONE CORRESPONDENCE BETWEEN THE ELEMENTS OF K AND L.

Let us consider a few examples. Suppose we have a half-line issuing from the point O, with a segment marked off indicating the unit of length. Further let Y be any point on the half-line. Then the segment OY can be measured, that is to say, one can correlate with it a certain non-negative number x called the length of the segment. Since this number depends exclusively on the position of the point Y, we may denote it by the symbol "$f(Y)$"; we consequently have:

$$x = f(Y).$$

But, conversely, to every non-negative number x, we may also construct a uniquely determined segment OY on the half-line under consideration, whose length equals x; in other words, to every x, there corresponds exactly one point Y such that

$$x = f(Y).$$

The function f is, therefore, biunique; it establishes a one-to-one correspondence between the points of the half-line and the non-negative numbers (and it would be equally simple to set up a one-to-one correspondence between the points of the entire line and all real numbers). Another example is supplied by the relation expressed by the formula:

$$x = -y.$$

This is a biunique function since, to every number x, there is only one number y satisfying the given formula; it can be seen at once that this function maps, for instance, the set of all positive numbers on the set of all negative numbers in a one-to-one manner. As a last example let us consider the relation expressed by the formula:

$$x = 2y$$

under the assumption that the symbol "y" here denotes natural numbers only. Again we have a biunique function; it correlates with every natural number y an even number $2y$; and vice versa—to every even natural number x there corresponds just one number y such that $2y = x$, namely, the number $y = \frac{1}{2}x$. The function thus establishes a one-to-one correspondence between arbitrary natural numbers and even natural numbers.—Numerous examples of biunique functions and one-to-one mappings can be drawn from the field of geometry (symmetric, collinear mappings, and so on).

Owing to the circumstance that we have the notion of a one-to-one correspondence at our disposal, we are now in a position to lay down an exact definition of a term which, earlier on, we had only been able to characterize intuitively rather than with precision. It is the concept of equinumerous classes (see Section 26). We shall now say that two classes K and L are equinumerous, or that they have the same cardinal number, if there exists a function which establishes a one-to-one correspondence between the elements of the two classes. On the basis of this definition it follows, in connection with the examples considered above, that the set of all points of an arbitrary half-line is equinumerous with the set of all non-negative numbers; and likewise, that the set of positive numbers and the set of negative numbers are equinumerous, and that the same holds for the set of all natural numbers and the set of all even natural numbers. The last example is particularly instructive; for it shows that a class may be equinumerous with a proper subclass of itself. To many readers this fact may seem most paradoxical at a first glance, because usually only finite classes are compared with respect to the numbers of their elements, and a finite class has, indeed, a greater cardinal number than any of its parts. The paradox disappears on calling to mind that the set of natural numbers is infinite and that we are, by no means, justified to ascribe properties to infinite classes that we have observed exclusively in connection with finite classes.—It is noteworthy that the property of the set of natural numbers of being equinumerous with one of its parts is shared by all infinite classes. This property is, therefore, characteristic of infinite classes, and it permits us to distinguish them from finite classes; a finite class can simply be defined as a class which is not equinumerous with any one of its proper subclasses. (However, this definition entails a certain logical difficulty, a discussion of which we will not enter into here.)[2]

34. Many-termed relations; functions of several variables and operations

We have, so far, considered exclusively TWO-TERMED (or BINARY) RELATIONS, that is, relations holding between two things. However, one also meets frequently with THREE-TERMED (or TERNARY) and, in general, MANY-TERMED RELATIONS within various

[2] The first to call attention to the property of infinite classes discussed here was the German philosopher and mathematician B. BOLZANO (1781–1848) in his book *Paradoxien des Unendlichen* (Leipzig 1851, posthumously published); in this work we already find the first beginnings of the contemporary theory of sets. The above property was later employed by PEIRCE (cf. footnote 2 on p. 14) and others in order to formulate an exact definition of a finite and of an infinite class.

sciences. In geometry, for instance, the relation of betweenness constitutes a typical example of a three-termed relation; it holds between three points of a line, and is expressed symbolically by the formula:

$$A/B/C$$

which is read:

the point B lies between the points A and C.

Arithmetic, too, supplies numerous examples of three-termed relations; it may suffice to mention the relation between three numbers x, y and z, consisting in the fact that the first number is the sum of the other two:

$$x = y + z,$$

as well as similar relations, such as are expressed by the following formulas:

$$x = y - z,$$

$$x = y \cdot z,$$

$$x = y : z.$$

As an example of a four-termed relation let us point to the relation holding between four points A, B, C and D if, and only if, the distance of the first two equals the distance of the last two, in other words, if the segments AB and CD are congruent. Another example is the relation holding between the numbers x, y, z and t whenever they form a proportion:

$$x : y = z : t.$$

Of particular importance among the totality of many-termed relations are the many-termed functional relations, which correspond to the two-termed functional relations. For reasons of simplicity we shall restrict ourselves to a discussion of three-termed relations of this type. R is called a THREE-TERMED FUNCTIONAL RELATION if, to any two things y and z, there corresponds at most one thing x having this relation to y and z. This uniquely determined thing, provided it exists at all, we denote either by the symbol:

$$R(y, z)$$

or else by the symbol:

$$y\,R\,z$$

(which now assumes a different meaning from what it had in the theory of two-termed relations). Thus, for the purpose of expressing that x stands to y and z in the functional relation R, we have two formulas at our disposal:

$$x = R(y, z) \quad \text{and} \quad x = y\,R\,z.$$

Corresponding to this twofold symbolism we have a twofold mode of expression. When using the notation:

$$x = R(y, z),$$

the relation R is called a FUNCTION. In order to differentiate between two-termed and three-termed functional relations, we speak, in the first case, of FUNCTIONS OF ONE VARIABLE or of FUNCTIONS WITH ONE ARGUMENT, and, in the second, of FUNCTIONS OF TWO VARIABLES or of FUNCTIONS WITH TWO ARGUMENTS. Similarly, four-termed functional relations are called FUNCTIONS OF THREE VARIABLES or FUNCTIONS WITH THREE ARGUMENTS, and so on. In designating functions with any number of arguments it is customary to employ the variables "f", "g", $\cdots$; the formula:

$$x = f(y, z)$$

is read:

x is that value of the function f which is correlated with the argument values y and z.

When the symbolism:

$$x = y\,R\,z$$

is employed, the relation R is usually referred to as an OPERATION or, more specifically, a BINARY OPERATION, and the above formula is read as follows:

x is the result of the operation R carried out on y and z;

in place of the letter "*R*" we tend to use, in this case, other letters, especially the letter "*O*". The four fundamental arithmetical operations of addition, subtraction, multiplication and division may serve as examples, and also such logical operations as addition and multiplication of classes or relations (see Sections 25 and 28). The content of the two concepts of a function of two variables and of a binary operation is evidently exactly the same. It should, perhaps, be noted that functions of one variable are sometimes also called operations, and, in particular, UNARY OPERATIONS; in the calculus of classes, for instance, the forming of the complement of a class is usually thought of, not as a function, but as an operation.

Although the many-termed relations play an important part in various sciences, the general theory of these relations is yet in its initial stage; when speaking of a relation, or of the theory of relations, one usually has only two-termed relations in mind. A more detailed study has so far only been made of one particular category of three-termed relations, namely, a category of binary operations, as the prototype of which we may consider the ordinary arithmetical addition. These investigations are carried on within the framework of a special mathematical discipline known as the theory of groups. We shall get acquainted with certain concepts from the theory of groups—and thereby also with certain general properties of binary operations—in the second part of this book.

35. The importance of logic for other sciences

We have discussed the most important concepts of contemporary logic, and in doing so we have got acquainted with some laws (very few, by the way) concerning these concepts. It had not been our intention, however, to give a complete list of all logical concepts and laws of which one avails oneself within scientific arguments. This, incidentally, is not necessary, as far as the study and promotion of other sciences are concerned, even of mathematics which is especially closely related to logic. Logic is justly considered the basis of all other sciences, if only for the reason that in every argument we employ concepts taken from the field of logic and that every correct inference proceeds in accordance with the laws of that discipline. But this does not imply that a thorough knowledge of logic is a necessary condition for correct thinking; even professional mathematicians, who, in general, do not commit errors in their inferences, usually do not know logic to such an extent as to be conscious of all logical laws of which they make unconscious use. All the same, there can be no doubt that the knowledge of logic is of considerable practical importance for everyone who desires to think and infer correctly, since it enhances the innate and acquired faculties to this effect and, in particularly critical cases, prevents the committing of mistakes. As far as, in particular, the construction of mathematical theories is concerned, logic plays a part of far-reaching importance also from the theoretical point of view; this problem will be discussed in the next chapter.

Exercises

1. Give examples of relations from the fields of arithmetic, geometry, physics, and everyday life.

2. Consider the relation of being father, that is to say, the relation expressed by the sentential function:

$$x \text{ is father of } y.$$

Do all human beings belong to the domain of this relation? And do they all belong to the counter-domain?

3. Consider the following seven relations among people, namely, of being father, mother, child, brother, sister, husband, wife. We denote these relations by the symbols "F", "M", "C", "B", "S", "H", "W". By applying various operations defined in Section 28 to the relations, we obtain new relations for which we sometimes find simple names in ordinary language; "H/C", for instance, as can be seen very easily, denotes the relation of being son-in-law. Find, if possible, simple names for the following relations:

$$\breve{B},\ \breve{H},\ H \cup W,\ F \cup B,\ F/M,\ M/\breve{C},\ B/\breve{C},\ F/(H \cup W),$$
$$(B/\breve{C}) \cup [H/(S/\breve{C})].$$

Express with the help of the symbols "F", "M", and so on, together with the symbols of the calculus of relations, the relations of being parent, sibling, grand-child, daughter-in-law and mother-in-law.

Explain the meanings of the following formulas, and determine which of them are true:

$$\mathbf{F} \subset \mathbf{M}', \quad \breve{\mathbf{B}} = \mathbf{S}, \quad \mathbf{F} \cup \mathbf{M} = \breve{\mathbf{C}}, \quad \mathbf{H}/\mathbf{M} = \mathbf{F}, \quad \mathbf{B}/\mathbf{S} \subset \mathbf{B},$$
$$\mathbf{S} \subset \mathbf{C}/\breve{\mathbf{C}}.$$

4. Consider the following two formulas of the calculus of relations:

$$R/S = S/R \quad \text{and} \quad \widetilde{(R/S)} = \breve{S}/\breve{R}.$$

Show by means of an example that the first is not always satisfied, and try to prove that the second is satisfied by arbitrary relations R and S.

Hint: Consider what it means to say that the relation $\widetilde{(R/S)}$ (that is, the converse of the relation R/S) or the relation $\breve{S}/\breve{R}$ holds between two things x and y.

5. Formulate in symbols the definitions of all terms of the calculus of relations that were discussed in Section 28.

Hint: The definition of the sum of two relations, for instance, has the following form:

$$[x\,(R \cup S)\,y] \leftrightarrow [(x\,R\,y) \vee (x\,S\,y)].$$

6. Which among the properties of relations discussed in Section 29 are possessed by the following relations:

(a) the relation of divisibility in the set of natural numbers;

(b) the relation of being relatively prime in the set of natural numbers (two natural numbers being called relatively prime if their greatest common divisor is 1);

(c) the relation of congruence in the set of polygons;

(d) the relation of being longer in the set of line segments;

(e) the relation of being perpendicular in the set of straight lines of a plane;

(f) the relation of intersecting in the set of geometric configurations;

(g) the relation of simultaneity in the class of physical events;

(h) the relation of temporally preceding in the class of physical events;

(i) the relation of being related in the class of human beings;

(k) the relation of fatherhood in the class of human beings?

7. Is it true that every relation is either reflexive or irreflexive (in the given class), and either symmetrical or asymmetrical? Give examples.

8. We shall call the relation R INTRANSITIVE IN THE CLASS K if, for any three elements x, y and z of K, the formulas:

$$x\,R\,y \quad \text{and} \quad y\,R\,z$$

imply the formula:

$$\sim(x\,R\,z).$$

Which of the relations listed in Exercises 3 and 6 are intransitive? Give other examples of intransitive relations. Is every relation either transitive or intransitive?

*9. Show how to make the transition from the expression:

the lines a and b are parallel

to the equivalent one:

the directions of the lines a and b are identical,

and how, in this connection, to define the expression "*the direction of a line*".

Solve the same problem for the following two expressions:

the segments AB and CD are congruent

and

the lengths of the segments AB and CD are equal.

What logical law has to be applied here?

Hint: Compare the remarks in Section 30 concerning the concept of similarity.

10. Let us agree to call two signs, or two expressions consisting of several signs, EQUIFORM, if they do not differ as far as their shape is concerned, but merely possibly with respect to their position in space, that is, with respect to the place at which they are printed; otherwise let us call them NON-EQUIFORM. For instance, in the formula:

$$x = x,$$

the variables on the two sides of the equality sign are equiform whereas we have non-equiform variables in the formula:

$$x = y.$$

Of how many signs does the formula:

$$x + y = y + x$$

consist? Into how many groups can these signs be divided, such that two equiform signs belong to the same group and two non-equiform signs belong to different groups?

Which of the properties discussed in Section 29 belong to the relations of equiformity and non-equiformity?

*11. Explain, on the basis of the results of the preceding exercise, why it may be said of equiform signs that they are equal with respect to their FORM, or that they have the same form, and how the term "*the form of the given sign*" is to be defined (compare Exercise 9).

It is a very common usage to call equiform signs simply equal and even to treat them as if they were one and the same sign. It is, for instance, often said that in an expression like:

$$x + x$$

one and the same variable occurs on both sides of the symbol "+". How should this be expressed with greater exactness?

*12. The inexact mode of speech which was pointed out in Exercise 11 has also been employed several times in this book (after all, we do not want to contend over deeply rooted usages).

Show that inexactitudes of this kind occur on pp. 12 and 56, and explain how they could be avoided.

Another example of an inexact mode of speech of this kind is the following: when speaking of sentential functions with one free variable one means functions in which all free variables are equiform. How can the expression:

sentential functions with two free variables

be formulated more exactly?

13. Given a point in a plane, consider the set of all circles in that plane with the given point as their common center. Show that this set is ordered by the relation of being a part. Would this be true too, if the circles did not lie in the same plane, or if they were not concentric?

14. We consider a relation among words of the English language which will be called the relation of PRECEDING (IN LEXICOGRAPHICAL ORDER). We shall explain here the meaning of this term by means of examples. The word "*and*" precedes the word "*can*", since the first begins with "*a*", the second with "*c*", and "*a*" has an earlier place in the English alphabet than "*c*". The word "*air*" precedes the word "*ale*", since they have the same first letter (or, rather, equiform first letters—cf. Exercise 10), while the second letter of the first word, that is "*i*", has an earlier place in the English alphabet than the second letter of the second word, that is "*l*". Analogously, "*each*" precedes "*eat*", and "*timber*" precedes "*time*". Finally, "*war*" precedes "*warfare*", since the first three letters of these words are the same, while the first word has only these letters, and the second more than these; and analogously "*mean*" precedes "*meander*".

Write the following words in a line so that, of any two words, the one on the left precedes the one on the right:

care, arm, salt, art, car, sale, trouble, army, ask.

Try to define the relation of preceding among words in a quite general way. Show that this relation establishes an order in the set of all English words. Point out some practical applications of this relation and explain why it is said to establish a lexicographical order.

15. Consider an arbitrary relation R and its negation R'. Show that the following statements of the theory of relations are true:

(a) *if the relation R is reflexive in the class K, then the relation R' is irreflexive in that class;*

(b) *if the relation R is symmetrical in the class K, then the relation R' is also symmetrical in that class K;*

*(c) *if the relation R is asymmetrical in the class K, then the relation R' is reflexive and connected in that class;*

*(d) *if the relation R is transitive and connected in the class K, then the relation R' is transitive in that class.*

Are the converses of these statements likewise true?

16. Show that, if the relation R has one of the properties discussed in Section 29, the converse relation $\breve{R}$ possesses the same property.

*17. The properties of relations which were introduced in Section 29 can easily be expressed in terms of the calculus of relations, provided the class K to which they refer is the universal class. The formulas:

$$R/R \subset R \quad \text{and} \quad \mathsf{D} \subset R \cup \breve{R},$$

for instance, express that the relation R is transitive and connected, respectively. Explain why; recall the meaning of the symbol "D" of Section 28. Express similarly that the relation R is symmetrical, asymmetrical, or intransitive (cf. Exercise 8). What property of relations discussed in the present chapter is expressed by the formula:

$$R/\breve{R} \subset \mathsf{I} \quad ?$$

18. Which of the relations expressed by the following formulas are functions:

(a) $2x + 3y = 12,$

(b) $x^2 = y^2,$

(c) $x + 2 > y - 3,$

(d) $x + y = y^2$,

(e) x *is mother of* y,

(f) x *is daughter of* y ?

Which of the relations considered in Exercise 3 are functions?

19. Consider the function expressed by the formula:

$$x = y^2 + 1.$$

What is the set of all argument values, and what is the set of all function values?

*20. Which of the functions in Exercise 18 are biunique? Give other examples of biunique functions.

*21. Consider the function expressed by the formula:

$$x = 3y + 1.$$

Show that this is a biunique function and that it maps the interval [0, 1] on the interval [1, 4] in a one-to-one manner (cf. Exercise 6 of Chapter IV). What conclusion may be drawn from this concerning the cardinal numbers of those intervals?

*22. Consider the function expressed by the formula:

$$x = 2^y.$$

Using this function show, along the lines of the preceding exercise, that the set of all numbers and the set of all positive numbers are equinumerous.

*23. Show that the set of all natural numbers and the set of all odd numbers are equinumerous.

24. Give examples of many-termed relations from the fields of arithmetic and geometry.

25. Which of the three-termed relations expressed by the following formulas are functions:

(a) $x + y + z = 0$,

(b) $x \cdot y > 2z$,

(c) $x^2 = y^2 + z^2$,

(d) $x + 2 = y^2 + z^2$?

26. Name a few laws of physics that state the existence of a functional relation between two, three and four quantities.

· VI ·

ON THE DEDUCTIVE METHOD

36. Fundamental constituents of a deductive theory—primitive and defined terms, axioms and theorems

We shall now attempt an exposition of the fundamental principles that are to be applied in the construction of logic and mathematics. The detailed analysis and critical evaluation of these principles are tasks of a special discipline, called the METHODOLOGY OF DEDUCTIVE SCIENCES or the METHODOLOGY OF MATHEMATICS. For anyone who intends to study or advance some science it is undoubtedly important to be conscious of the method which is employed in the construction of that science; and we shall see that, in the case of mathematics, the knowledge of that method is of particularly far-reaching importance, for lacking such knowledge it is impossible to comprehend the nature of mathematics.

The principles with which we shall get acquainted serve the purpose of securing for the knowledge acquired in logic and mathematics the highest possible degree of clarity and certainty. From this point of view a method of procedure would be ideal, if it permitted us to explain the meaning of every expression occurring in this science and to justify each of its assertions. It is easy to see that this ideal can never be realized. In fact, when one tries to explain the meaning of an expression, one uses, of necessity, other expressions; and in order to explain, in turn, the meaning of these expressions, without entering into a vicious circle, one has to resort to further expressions again, and so on. We thus have the beginning of a process which can never be brought to an end, a process which, figuratively speaking, may be characterized as an

INFINITE REGRESS—a *regressus in infinitum.* The situation is quite analogous as far as the justification of the asserted statements of the science is concerned; for, in order to establish the validity of a statement, it is necessary to refer back to other statements, and (if no vicious circle is to occur) this leads again to an infinite regress.

By way of a compromise between that unattainable ideal and the realizable possibilities, certain principles concerning the construction of mathematical disciplines have emerged that may be described as follows.

When we set out to construct a given discipline, we distinguish, first of all, a certain small group of expressions of this discipline that seem to us to be immediately understandable; the expressions of this group we call PRIMITIVE TERMS or UNDEFINED TERMS, and we employ them without explaining their meanings. At the same time we adopt the principle: not to employ any of the other expressions of the discipline under consideration, unless its meaning has first been determined with the help of primitive terms and of such expressions of the discipline whose meanings have been explained previously. **The sentence which determines the meaning of a term in this way is called a DEFINITION, and the expressions themselves whose meanings have thereby been determined are accordingly known as DEFINED TERMS.**

We proceed similarly with respect to the asserted statements of the discipline under consideration. Some of these statements which to us have the appearance of evidence are chosen as the so-called PRIMITIVE STATEMENTS or AXIOMS (also often referred to as POSTULATES, but we shall not use the latter term in this technical meaning here); we accept them as true without in any way establishing their validity. On the other hand, we agree to accept any other statement as true only if we have succeeded in establishing its validity, and to use, while doing so, nothing but axioms, definitions and such statements of the discipline the validity of which has been established previously. As is well known, statements established in this way are called PROVED STATEMENTS or THEOREMS, and the process of establishing them is called a PROOF. More generally, if within logic or mathematics we establish one statement on the basis of others, we refer to this process as a

DERIVATION OR DEDUCTION, and the statement established in this way is said to be DERIVED OR DEDUCED from the other statements or to be their CONSEQUENCE.

Contemporary mathematical logic is one of those disciplines which are constructed in accordance with the principles just stated; unfortunately, it has not been possible within the narrow framework of this book to give this important fact due prominence. If any other discipline is constructed according to these principles, it is already based upon logic; logic, so to speak, is then already presupposed. This means that all expressions and laws of logic are treated on an equal footing with the primitive terms and axioms of the discipline under construction; the logical terms are used in the formulation of the axioms, theorems and definitions, for instance, without an explanation of their meaning, and the logical laws are applied in proofs without first establishing their validity. Sometimes it is even convenient not only to use logic in the construction of a discipline but to presuppose in the same sense certain mathematical disciplines previously constructed; for reasons of brevity, these theories, together with logic, may be characterized as the DISCIPLINES PRECEDING THE GIVEN DISCIPLINE. Thus logic itself does not presuppose any preceding discipline; in the construction of arithmetic as a special mathematical discipline logic is presupposed as the only preceding discipline; on the other hand, in the case of geometry it is expedient—though not unavoidable—to presuppose not only logic but also arithmetic.

With reference to the last remarks it is necessary to make certain corrections in the formulation of the principles stated above. Before undertaking the construction of a discipline those disciplines have to be enumerated that are to precede the given discipline; all requirements concerning the defining of expressions and the proving of statements, however, are limited to those expressions and statements which are specific for the discipline under construction, that is those which do not belong to the preceding disciplines.

The method of constructing a discipline in strict accordance with the principles laid down above is known as the DEDUCTIVE

METHOD; and the disciplines constructed in this manner are called DEDUCTIVE THEORIES.[1] The view has become more and more common that the deductive method is the only essential feature by means of which the mathematical disciplines can be distinguished from all other sciences; not only is every mathematical discipline a deductive theory, but also, conversely, every deductive theory is a mathematical discipline (according to this view deductive logic is also to be counted among the mathematical disciplines). We will not enter here into a discussion of the reasons in favor of this view, but merely remark that it is possible to put forward ponderable arguments in its support.

37. Model and interpretation of a deductive theory

As a result of a consistent application of the principles presented in the preceding section, deductive theories acquire certain interesting and important features which we shall describe here. Since the questions which we are going to discuss have a rather involved and abstract character, we shall try to elucidate them by means of a concrete example.

Suppose we are interested in general facts about the congruence of line segments, and we intend to build up this fragment of geometry as a special deductive theory. We accordingly stipulate that the variables "x", "y", "z", . . . denote segments. As primitive terms we choose the symbols "S" and "$\cong$". The former is

[1] The deductive method cannot be considered an achievement of recent times. Already in the *Elements* of the Greek mathematician EUCLID (about 300 B.C.) we find a presentation of geometry which leaves nothing much to be desired from the standpoint of the methodological principles stated above. For 2200 years, mathematicians have seen in EUCLID'S work the ideal and prototype of scientific exactitude. An essential progress in this field occurred only during the last 50 years, in the course of which the foundations of the basic mathematical disciplines of geometry and arithmetic were laid in accordance with all requirements of the present-day methodology of mathematics. Among the works to which we are indebted for this progress we will mention at least the following two, which have already become of historic importance: the collective work *Formulaire de Mathématiques* (Torino 1895–1908) whose editor and main author was the Italian mathematician and logician G. PEANO (1858–1932), and *Grundlagen der Geometrie* (Leipzig and Berlin 1899) by the great contemporary German mathematician D. HILBERT.

an abbreviation of the term "*the set of all segments*"; the latter designates the relation of congruence, so that the formula:

$$x \cong y$$

is to be read as follows:

the segments x and y are congruent.

Further we adopt only two axioms:

AXIOM I. *For any element x of the set* S, $x \cong x$ (in other words: *every segment is congruent to itself*).

AXIOM II. *For any elements x, y and z of the set* S, *if* $x \cong z$ *and* $y \cong z$, *then* $x \cong y$ (in other words: *two segments congruent to the same segment are congruent to each other*).

Various theorems on the congruence of segments may be derived from these axioms, for instance:

THEOREM I. *For any elements y and z of the set* S, *if* $y \cong z$, *then* $z \cong y$.

THEOREM II. *For any elements x, y and z of the set* S, *if* $x \cong y$ *and* $y \cong z$, *then* $x \cong z$.

The proofs of these two theorems are very easy. Let us, for instance, sketch the proof of the first.

Putting in Axiom II "z" for "x" we obtain:

for any elements y and z of the set S, *if* $z \cong z$ *and* $y \cong z$, *then* $z \cong y$.

In the hypothesis of this statement we have the formula:

$$z \cong z$$

which, on the basis of Axiom I, is undoubtedly valid, and may hence be omitted. We thus arrive at the theorem in question.

In connection with these simple considerations we want to make the following remarks.

Our miniature deductive theory rests upon a suitably selected system of primitive terms and axioms. Our knowledge of the things denoted by the primitive terms, that is, of the segments and their congruence, is very comprehensive and is by no means exhausted by the adopted axioms. But this knowledge is, so to speak, our private concern which does not exert the least influence on the construction of our theory. In particular, in deriving theorems from the axioms, we make no use whatsoever of this knowledge, and behave as though we did not understand the content of the concepts involved in our considerations, and as if we knew nothing about them that had not been expressly asserted in the axioms. We disregard, as it is commonly put, the meaning of the primitive terms adopted by us, and direct our attention exclusively to the form of the axioms in which these terms occur.

This implies a very significant and interesting consequence. Let us replace the primitive terms in all axioms and theorems of our theory by suitable variables, for instance, the symbol "**S**" by the variable "*K*" denoting classes, and the symbol "$\cong$" by the variable "*R*" denoting relations (in order to simplify the considerations, we disregard here any theorems which contain defined terms). The statements of our theory will then be no longer sentences, but will become sentential functions which contain two free variables, "*K*" and "*R*", and which express, in general, the fact that the relation *R* has this or that property in the class *K* (or, more precisely, that this or that relation holds between *K* and *R*; cf. Section 27). For instance, as it is easily seen, Axiom I and Theorems I and II will now say that the relation *R* is reflexive, symmetrical and transitive respectively, in the class *K*. Axiom II will express a property for which we do not have any special name and to which we shall refer as property **P**; this is the following property:

for any elements x, y and z of the class K, if xRz and yRz, then xRy.

Since, in the proofs of our theory, we make use of no properties of the class of segments and of the relation of congruence but those which were explicitly stated in the axioms, every proof can be considerably generalized, for it can be applied to any class *K* and any relation *R* having those properties. As a result of such a generalization of the proofs, we can correlate with any theorem of our theory a general law belonging to the domain of logic, namely to the theory of relations, and stating that every relation R which is reflexive and has the property **P** in the class K also has the property expressed in the theorem considered. So, for instance, the following two laws of the theory of relations correspond to Theorems I and II:

I′. *Every relation R which is reflexive in the class K and has the property* **P** *in that class is also symmetrical in K.*

II′. *Every relation R which is reflexive in the class K and has the property* **P** *in that class is also transitive in K.*

If a relation R is reflexive and has the property **P** in a class K, we say that K and R together form a MODEL OR A REALIZATION OF THE AXIOM SYSTEM of our theory, or, simply, that they satisfy the axioms. One model of the axiom system is formed, for instance, by the class of the segments and the relation of congruence, that is, the things denoted by the primitive terms; of course, this model also satisfies all the theorems deduced from the axioms. (To be exact, we ought to say that a model satisfies not the statements of the theory themselves, but the sentential functions obtained from them by replacing the primitive terms by variables.) However, this particular model does not play any privileged role in the construction of the theory. On the contrary, on the basis of universal logical laws like I′ and II′ we arrive at the general conclusion that any model of the axiom system satisfies all theorems deduced from these axioms. In view of this fact, a model of the axiom system of our theory is also referred to as a MODEL OF THE THEORY itself.

We are able to exhibit many different models for our axiom system, even in the domain of logic and mathematics. To obtain such a model, we select within any other deductive theory two constants, say "**K**" and "**R**" (the former denoting a class, the latter a relation), then we replace "**S**" by "**K**" and "≅" by "**R**" everywhere in the system, and finally we show that the sentences thus obtained are theorems, or possibly axioms, of the new theory. If we have succeeded in doing so, we say that we have found an INTERPRETATION OF THE AXIOM SYSTEM—and, at the same time, of our whole DEDUCTIVE THEORY—WITHIN THE OTHER DEDUCTIVE THEORY. If we now replace the primitive terms "**S**" and "≅"

by "**K**" and "**R**", not only in the axioms, but also in all theorems of our theory, we can be sure in advance that all sentences thus obtained will be true sentences of the new deductive theory.

We shall give here two concrete examples of interpretations of our miniature theory. Let us replace in Axioms I and II the symbol "**S**" by the symbol of the universal class "V", and the symbol "$\cong$" by the identity sign "$=$". As can be seen immediately, the axioms will then become logical laws (in fact, Laws II and V of Section 17 in a slightly modified form). The universal class and the relation of identity constitute, therefore, a model of the axiom system, and our theory has found an interpretation within logic. Thus, if in Theorems I and II we replace the symbols "**S**" and "$\cong$" by the symbols "V" and "$=$", we are sure to arrive at true logical sentences (in fact, we are again familiar with them—cf. Laws III and IV of Section 17).

Next, let us consider the set of all numbers, or any other set of numbers, denoting it by "**N**". Let us call two numbers x and y equivalent, in symbols:

$$x \equiv y,$$

if their difference $x - y$ is an integer; thus we have, for example:

$$1\tfrac{1}{4} \equiv 5\tfrac{1}{4},$$

whereas it is not the case that

$$3 \equiv 2\tfrac{1}{3}.$$

If now, in both axioms, the primitive terms are replaced by "**N**" and "$\equiv$", it can be easily shown that the resulting sentences are true theorems of arithmetic. Thus our theory possesses an interpretation within arithmetic, for the set of numbers **N** and the relation of equivalence $\equiv$ constitute a model of the axiom system. And again, without any special reasoning we are sure that Theorems I and II will become true arithmetical statements if they are subjected to the same transformation as the axioms.

The general facts described above have many interesting applications in methodological researches. We shall illustrate this here by means of one example only; we shall show how it may be proved —on the basis of these facts—that certain sentences cannot be deduced from our axiom system.

Let us consider the following sentence A (formulated in logical terms and in the primitive terms of our theory only):

A. *There exist two elements x and y of the set* S *for which it is not the case that* $x \cong y$ (in other words: *there exist two segments which are not congruent*).

This sentence seems to be undoubtedly true. Nevertheless, no attempts to prove it on the basis of Axioms I and II give a positive result. Thus the conjecture arises that Sentence A cannot be deduced at all from our axioms. In order to confirm this conjecture, we argue in the following way. If Sentence A could be proved on the basis of our axiom system, then, as we know, every model of this system would satisfy that sentence; if, therefore, we succeed in indicating such a model of the axiom system which will not satisfy Sentence A, we shall prove thereby that this sentence cannot be deduced from Axioms I and II. Now, it turns out that producing such a model does not present any difficulties. Let us consider, for instance, the set of all integers I (or any other set of integers, e.g. the set consisting of the numbers 0 and 1 only) and the relation of equivalence $\equiv$ between numbers which was discussed above. We already know from the preceding remarks that the set I and the relation $\equiv$ constitute a model of our axiom system; Sentence A however is not satisfied by this model, for there are no two integers x and y which are not equivalent, that is, whose difference is not an integer. Another model appropriate to this purpose is formed by an arbitrary class of individuals and by the universal relation V holding between any two individuals.

The type of reasoning just applied is known as the METHOD OF PROOF BY EXHIBITING A MODEL OR BY INTERPRETATION.

The facts and concepts discussed here can be related, without essential change, to other deductive theories. In the next section we shall try to describe them in a quite general way.

38. Law of deduction; formal character of deductive sciences

*We consider any deductive theory based upon a system of primitive terms and axioms. In order to simplify our considerations, we assume that this theory presupposes logic only, that is, logic is the only theory preceding the given theory (cf. Section 36).

Let us imagine that in all the statements of our theory the primitive terms are replaced by suitable variables throughout (as in Section 37, and again for the sake of simplicity, we disregard theorems containing defined terms). The statements of the theory considered become sentential functions containing as free variables those symbols by which the primitive terms had been replaced and not containing any constants but those belonging to logic. Given certain things one can find out whether they satisfy all the axioms of our theory, or, to be exact, the sentential functions obtained from these axioms in the manner just described (that is, whether the names or designations of those things, when put in the place of the free variables, render the sentential functions true sentences; cf. Section 2). If it turns out that this is the case, we shall say that the things under consideration form a MODEL or a REALIZATION OF THE AXIOM SYSTEM of our deductive theory; we also say sometimes that they form a MODEL OF THE DEDUCTIVE THEORY itself. In a quite analogous manner we can find out whether given things satisfy not only the axiom system but also any other system of statements of our theory and whether, therefore, they form a model of this system (it is not excluded that the system consists of a single statement).

A model of the axiom system is formed, for instance, by those things which are denoted by the primitive terms of the given theory, since we assume that all axioms are true sentences; this model satisfies, of course, all the theorems of our theory. But as far as the construction of our theory is concerned, this model takes no distinguished place among all the other models. When deducing this or that theorem from the axioms, we do not think of the specific properties of this model, and we make use of only those properties which are explicitly stated in the axioms and, therefore, belong to every model of the axiom system. Consequently, every proof of a particular theorem of our theory can be extended to every model of the axiom system and can be thus transformed into a much more general argument no longer belonging to our theory but to logic; and as a result of this generalization we obtain a general logical statement (like the laws I′ and II′ of the preceding section) which establishes the fact that the theorem in question is satisfied by every model of our axiom system.

The final conclusion at which we arrive in this way can be put in the following form:

Every theorem of a given deductive theory is satisfied by any model of the axiom system of this theory; and moreover, to every theorem there corresponds a general statement which can be formulated and proved within the framework of logic and which establishes the fact that the theorem in question is satisfied by any such model.

We have here a general law from the domain of the methodology of deductive sciences which, when formulated in a slightly more precise way, is known as the LAW OF DEDUCTION (or the DEDUCTION THEOREM).[2]

The tremendous practical importance of this law results from the fact that we are usually able to exhibit numerous models of the axiom system of a particular theory, even without leaving the field of the deductive sciences. In order to arrive at such a model it is sufficient to select certain constants from some other deductive theory (which can be logic or a theory presupposing logic), to put them in the axioms in place of the primitive terms, and to show that the sentences obtained in this way are asserted statements of that other theory. We say in this case that we have found an INTERPRETATION OF THE AXIOM SYSTEM OF THE ORIGINAL THEORY WITHIN THE OTHER THEORY. (It may, in particular, occur that the constants chosen belong to the theory originally considered, in which case some of the primitive terms may even have remained unchanged; the given axiom system is then said to have found a new interpretation within the theory under consideration.) We shall also subject the theorems of the original theory to an analogous transformation, replacing the primitive terms throughout by those constants that had been employed in the interpretation of the axioms. On the basis of the law of deduction we can then be sure in advance that the sentences arrived at in this manner are asserted statements of the new theory. We can formulate this in the following way:

[2] This law was formulated by the author as a general methodological postulate, and was later proved exactly for various particular deductive theories.

All theorems proved on the basis of a given axiom system remain valid for any interpretation of the system.

It is redundant to give a special proof for any of these transformed theorems; it would in any case be a task of a purely mechanical nature, for it would be sufficient to transfer the corresponding argument from the field of the original theory and to subject it to the same transformations that had been carried out with respect to the axioms and theorems. Every proof within a deductive theory contains—potentially, so to speak,—an unlimited number of other analogous proofs.

The facts described above demonstrate the great value of the deductive method from the point of view of economy of human thought. They are also of far-reaching theoretical importance, if only for this reason that they establish a foundation for various arguments and researches within the methodology of deductive sciences. In particular, the law of deduction is the theoretical basis for all so-called PROOFS BY INTERPRETATION; we have already encountered one example of such proofs in the preceding section, and we shall meet with various other examples in the second part of this book.

For reasons of exactness it may be added that the considerations sketched here are applicable to any deductive theory in whose construction logic is presupposed, whereas their application to logic itself brings about certain difficulties which we would rather not discuss here. If a deductive theory presupposes not only logic, but also other theories, some of the formulations given above assume a somewhat more complicated form.

The common source of the methodological phenomena discussed here is the fact pointed out in the preceding section, namely that, in constructing a deductive theory, we disregard the meaning of the axioms and take into account only their form. It is for this reason that people when referring to those phenomena speak about the purely FORMAL CHARACTER of deductive sciences and of all reasonings within these sciences.

From time to time one finds statements which emphasize the formal character of mathematics in a paradoxical and exaggerated way; although fundamentally correct, these statements may become a source of obscurity and confusion. Thus one hears and even reads occasionally that no definite content may be ascribed to mathematical concepts; that in mathematics we do not really know what we are talking about, and that we are not interested in whether our assertions are true. One should approach such judgments rather critically. If, in the construction of a theory, one behaves as if one did not understand the meaning of the terms of this discipline, this is not at all the same as denying those terms any meaning. It is, admittedly, sometimes the case that we develop a deductive theory without ascribing a definite meaning to its primitive terms, thus dealing with the latter as with variables; in this case we say that we treat the theory as a FORMAL SYSTEM. But this is a comparatively rare situation (not even taken into account in our general characterization of deductive theories given in Section 36), and it occurs only if it is possible to give several interpretations for the axiom system of this theory, that is, if there are several ways available of ascribing concrete meanings to the terms occurring in the theory, but if we do not desire to give preference in advance to any one of these ways. A formal system, on the other hand, for which we are unable to give a single interpretation, would, presumably, be of interest to nobody.

In conclusion we shall call attention to certain interesting examples of interpretations of mathematical disciplines, which are much more important than those given in Section 37.

The axiom system of arithmetic may be interpreted within geometry: given an arbitrary straight line, it is possible to define such relations between its points and operations on its points as satisfy all the axioms—and hence also all the theorems—of arithmetic, which are concerned with corresponding relations between numbers and operations on numbers. (This is closely connected with a circumstance mentioned in Section 33, namely, the possibility of establishing a one-to-one correspondence between all the points of a line and all numbers.) Conversely, the axiom system of geometry also possesses an interpretation within arithmetic. The uses to which these two facts can be put are manifold. Geometrical configurations may, for instance, be employed in order to give a visual image of various facts in the field of arithmetic,—a procedure known as the graphical method; on the other hand, it is possible to investigate geometrical facts with the help of arithmetical or algebraical methods,—there is even a special branch of geometry, known as analytic geometry, which is concerned with all investigations of this type.

Arithmetic, as we have seen previously, may be built up as a part of logic (cf. Section 26). But if we treat arithmetic as an independent deductive theory, resting upon its own system of primitive terms and axioms, its relation to logic can be described as follows: arithmetic possesses an interpretation within logic (with the understanding that the axiom of infinity be included in logic,—cf. Section 26); in other words, it is possible within logic to define such concepts as satisfy all the axioms, and hence also all the theorems, of arithmetic. If we remember that geometry has an interpretation in arithmetic, we arrive at the conclusion that also geometry can be interpreted within logic. All these are facts which are exceedingly significant from the methodological point of view.*

39. Selection of axioms and primitive terms; their independence

We will now turn to the discussion of a few problems of a more special nature, which, however, concern fundamental components of the deductive method, namely the choice of the primitive terms and axioms as well as the construction of definitions and proofs.

It is important to realize the fact that we have a large degree of freedom in the selection of the primitive terms and axioms; it would be quite erroneous to believe that certain expressions cannot be defined in any possible way, or that certain statements can, on principle, not be proved. Let us call two systems of sentences of a given theory EQUIPOLLENT, if each sentence of the first system can be derived from the sentences of the second, together with theorems of the preceding theories, and, conversely, if every sentence of the second system can be derived from the sentences of the first (if any sentences occur in both systems, they do not, of course, have to be derived). Let us imagine, further, that a certain deductive theory has been built up on the basis of some axiom system, and that in the course of its construction we come across a system of statements equipollent in the sense just defined to the axiom system. (A concrete example can be obtained in connection with the miniature theory of the congruence of segments discussed in Section 37: it is easy to show that its axiom system is equipollent to the system of sentences consisting of Axiom I together with Theorems I and II.) If this kind of situation arises, then, from the theoretical point of view, it would be possible to reconstruct the entire theory in such a manner that the statements of the new system are taken as axioms, while the former axioms are proved as theorems. Even the circumstance that the new axioms may, at first, to a much lesser degree have the appearance of immediate evidence is inessential; for every sentence becomes evident to a certain degree, once it has been derived in a convincing manner from other evident sentences. All this applies likewise—*mutatis mutandis*—to the primitive terms of a deductive theory; the system of these terms may be replaced by any other system of terms of the theory in question, provided only the two systems are EQUIPOLLENT in the sense that each term of the first system can be defined by means of terms of the second together with terms taken from the preceding theories, and vice versa. It is not for theoretical reasons (or, at least, not only for theoretical reasons) that we decide to select a certain system of primitive terms and axioms in preference to any of the other possible equipollent systems; other factors play a role here,—practical, didactical, even esthetic ones. Sometimes it is a question of choosing the simplest possible primitive terms and axioms, then again it may be desirable to get along with as few of them as possible, or we may prefer such primitive terms and axioms as would enable us, in the simplest possible way, to define those terms and to prove those statements of a given theory in which we are especially interested.

Another problem arises in close connection with these remarks. Fundamentally, we strive to arrive at an axiom system which does not contain a single superfluous statement, that is, a statement which can be derived from the remaining axioms and which, therefore, might be counted among the theorems of the theory under construction. An axiom system of this kind is called INDEPENDENT (or a SYSTEM OF MUTUALLY INDEPENDENT AXIOMS). We likewise attempt to see to it that the system of primitive terms is INDEPENDENT, that is, that it does not contain any superfluous term which can be defined by means of the others. Often, however, one does not insist on these methodological postulates for practical, didactical reasons, particularly in cases where the omission of a superfluous axiom or primitive term would bring about great complications in the construction of the theory.

40. Formalization of definitions and proofs, formalized deductive theories

The deductive method is justifiably considered the most perfect of all methods employed in the construction of sciences. It disposes to a large extent of the possibility of obscurities and errors, without resorting to an infinite regress; and it is due to this method that any reasons for doubt as to the content of concepts or the truth of assertions of a given theory are considerably reduced and may hold at most for the few primitive terms and axioms.

One reservation has to be added to this statement however. The application of the deductive method will give the desired results only if all the definitions and proofs fulfil their tasks completely, that is, if the definitions make fully clear the meaning of all the terms to be defined and if the proofs convince us wholly of the validity of all the theorems to be proved. It is far from easy to examine whether the definitions and proofs actually comply with these requirements; it is quite possible, for instance, that an argument which seems entirely convincing to one person is not even comprehensible to another. In order to remove any cause for doubt in this respect, the present-day methodology endeavors to replace subjective valuations in the examination of definitions and proofs by criteria of an objective nature, and to make the decision as to the correctness of definitions or proofs dependent exclusively upon their structure, that is, their exterior form. For this purpose, special RULES OF DEFINITION and RULES OF PROOF (or OF INFERENCE) are stated. The first tell us what form the sentences should have which are used as definitions in the theory under consideration, and the second describe the kind of transformations to which statements of this theory may be subjected in order to derive other statements from them; each definition has to be laid down in accordance with the rules of definition, and each proof must be COMPLETE, that is, it must consist in a successive application of rules of proof to sentences previously recognized as true (cf. Sections 11 and 15).—These new methodological postulates may be denoted as postulates of the FORMALIZATION OF DEFINITIONS AND PROOFS; a discipline constructed in accordance with these new postulates is called a FORMALIZED DEDUCTIVE THEORY.[3]

Through the postulates of formalization the formal character of mathematics is enhanced considerably. Already at an earlier stage in the development of the deductive method we were, in the construction of a mathematical discipline, supposed to disregard the meanings of all expressions specific to this discipline, and we were to behave as if the places of these expressions were taken by variables void of any independent meaning. But, at least, to the logical concepts we were permitted to ascribe their customary meanings. In this connection, the axioms and theorems of a mathematical discipline could be treated, if not as sentences, then at least as sentential functions, that is, as expressions having the grammatical form of sentences and expressing certain properties of things or relations among things. To derive a theorem from accepted axioms (or from theorems previously proved) was the same as to show convincingly that all things satisfying the axioms also satisfy the theorem in question; mathematical proofs did not altogether differ very much from considerations of everyday life. Now, however, the meanings of all expressions encountered in the given discipline are to be disregarded without exception, and we are supposed to behave in the task of constructing a deductive theory as if its sentences were configurations of signs void of any content; each proof will now consist in subjecting axioms or previously proved theorems to a series of purely external transformations.

[3] The first attempts to present the deductive theories in a formalized form are due to FREGE who has already been quoted twice (cf. footnote 2 on p. 19). A very high level in the process of formalization was achieved in the works of the late Polish logician S. LEŚNIEWSKI (1886–1939); one of his achievements is an exact and exhaustive formulation of the rules of definition.

In the light of modern requirements, logic becomes the basis of the mathematical sciences in a much more thorough sense than it used to be. We may no longer be satisfied with the conviction that—due to our innate or acquired capacity for correct thinking—our argumentations are in accordance with the rules of logic. In order to give a complete proof of a theorem it is necessary to apply the transformations prescribed by the rules of proof not only to the statements of the theory with which we are concerned, but also to those of logic (and other preceding theories); and for this purpose we have to have a complete list of all logical laws at our disposal that are applied in the proofs.

It is only by virtue of the development of deductive logic that, theoretically at least, we are today in a position to present every mathematical discipline in formalized form. In practice, however, this still involves considerable complications; a gain in exactitude and methodological correctness is accompanied by a loss in clarity and intelligibility. The whole problem, after all, is fairly new, the relevant investigations are not yet definitely concluded, and there is reason to hope that their further pursuance will eventually bring about essential simplifications. It would therefore be premature to comply fully at the present time, in a popular presentation of any part of mathematics, with the postulates of formalization. In particular, it would be far from sensible to demand that the proofs of theorems in an ordinary textbook of some mathematical discipline be given in complete form; one should, however, expect the author of a textbook to be intuitively certain that all his proofs can be brought into that form, and even to carry his considerations to a point from which a reader who has some practice in deductive thinking and sufficient knowledge of contemporary logic would be able to fill the remaining gaps without much difficulty.

41. Consistency and completeness of a deductive theory; decision problem

We shall now consider two methodological concepts which are of great importance from the theoretical point of view, while in practical respects they are of little significance. They are the concepts of CONSISTENCY and of COMPLETENESS.

A deductive theory is called CONSISTENT or NON-CONTRADICTORY if no two asserted statements of this theory contradict each other, or, in other words, if of any two contradictory sentences (cf. Section 7) at least one cannot be proved. A theory is called COMPLETE, on the other hand, if of any two contradictory sentences formulated exclusively in the terms of the theory under consideration (and the theories preceding it) at least one sentence can be proved in this theory. Of a sentence which has the property that its negation can be proved in a given theory, it is usually said that it can be DISPROVED in that theory. In this terminology we can say that a deductive theory is consistent if no sentence can be both proved and disproved in it; a theory is complete, on the other hand, if every sentence formulated in the terms of this theory can be proved or disproved in it. Both terms "consistent" and "complete" are applied, not only to the theory itself, but also to the axiom system upon which it is based.

Let us now try to get a clear idea of the import of these two notions. Every discipline, even one constructed entirely correctly in every methodological respect, loses its value in our eyes if we have reason to suspect that not all assertions of this discipline are true. On the other hand, the value of a discipline will be the greater, the larger the number of true sentences whose validity can be established in it. From this point of view, a discipline might be considered ideal, if it contains among its asserted statements all true sentences which are relevant to that theory, and not a single false one. A sentence is here considered relevant if it is formulated entirely in terms of the discipline under consideration (and its preceding disciplines); after all, it cannot be expected that, say, in arithmetic all true sentences can be proved, even such as contain concepts of chemistry or biology.—Let us now imagine that a deductive theory is inconsistent, that is to say, that two contradictory sentences occur among its axioms and theorems; from a well-known logical law, namely, the law of contradiction (cf. Section 13), it follows that one of these sentences must be false. If, on the other hand, we assume the theory to be incomplete, there exist two relevant contradictory sentences of which neither can be proved in that discipline; and yet, by another logical law, i. e., the law of excluded middle, one of the two sentences must be true. We see from this that a deductive theory certainly falls short of our ideal unless it is both consistent and complete. (Thereby we do not mean to imply that every consistent and complete discipline must, *ipso facto*, be a realization of our ideal, that is, that it must contain among its asserted statements all true sentences and only such sentences.)

There is yet another aspect to the whole question which we have been considering. The development of any deductive science consists in formulating in the terms of this science problems of the type *"is such and such the case?"* and then attempting to decide these problems on the basis of the axioms that have been assumed. Any problem of this type may clearly be decided in one of two possible ways: in the affirmative or in the negative. On the first alternative, the answer runs: *"such and such is the case"*; and on the second: *"such and such is not the case"*. The consistency and the completeness of the axiom system of a deductive theory now give us a guarantee that every problem of the kind mentioned can actually be decided within the theory, and moreover decided in one way only; the consistency excludes the possibility that any problem may be decided in two ways, that is, both affirmatively and negatively, and the completeness assures us that it can be decided in at least one way.

Closely connected with the problem of completeness is another, more general, problem which concerns incomplete as well as complete theories. It is the problem which consists in finding, for the given deductive theory, a general method which would enable us to decide whether or not any particular sentence formulated in the terms of this theory can be proved within this theory. This important problem is known as the DECISION PROBLEM.[4]

There are only a few deductive theories known of which it has been possible to show that they are consistent and complete.

[4] The import of the concepts and problems discussed in this section—and especially of the concept of consistency and of the decision problem—was emphasized by HILBERT (cf. footnote 1 on p. 120), who greatly stimulated many important investigations into the foundations of mathematics. Upon his instigation, these concepts and problems have of late been made the subject of intensive researches by a number of contemporary mathematicians and logicians.

They are, as a rule, elementary theories of a simple logical structure and a modest stock of concepts. An example is given by sentential calculus, which has been discussed in Chapter II, provided that it is considered as an independent theory and not as a part of logic (however, in applying the term "complete" to this theory, it is to be used in a slightly modified meaning). Perhaps the most interesting example of a consistent and complete theory is that supplied by elementary geometry; we have here in mind geometry limited to those confines wherein it has for centuries been taught in schools as a part of elementary mathematics, that is to say, a discipline in which the properties of various special kinds of geometrical figures such as lines, planes, triangles, circles are investigated, but in which the general concept of a geometrical configuration (a point set) does not occur.[5] The situation changes essentially as soon as one goes over to such sciences as arithmetic or advanced geometry. Probably no one working in these sciences doubts their consistency; and yet, as has resulted from the latest methodological investigations, a strict proof of their consistency meets with great difficulties of a fundamental nature. The situation in regard to the problem of completeness is even worse; it turns out that arithmetic and advanced geometry are incomplete; for it has been possible to set up problems of a purely arithmetical or geometrical character that can be neither positively nor negatively decided within these disciplines. It might be supposed that this fact is merely an outcome of the imperfection of the axiom systems and methods of proof at our disposal up to date, and that a suitable modification (for instance, an extension of the axiom system) may, in the future, yield complete systems. Deeper investigations, however, have shown this conjecture to be erroneous: never will it be possible to build up a consistent and complete deductive theory containing as its theorems all true sentences of arithmetic or of advanced geometry. Moreover, it turns out that the decision problem likewise does not admit of a

[5] For the first proof of completeness of sentential calculus (and, thereby, for the first positive result of the investigations concerned with completeness) we are indebted to the contemporary American logician E. L. Post. The proof of the completeness of elementary geometry originates with the author.

positive solution with respect to these disciplines; it is impossible to set up a general method which would allow us to differentiate between those sentences which can be proved within these disciplines and those which cannot be proved. All these results can be extended to many other deductive theories, and, in particular, to all those which either presuppose the arithmetic of integers (i.e., the theory of the four basic arithmetical operations on integers) or contain sufficient devices to develop this theory. *Thus, for instance, these results can be applied to the general theory of classes (as follows from the remarks at the end of Section 26).*[6]

In view of these last remarks, it is understandable that the concepts of consistency and completeness—in spite of their theoretical importance—exert little influence in practice upon the construction of deductive theories.

42. The widened conception of the methodology of deductive sciences

The investigations concerning consistency and completeness were among the most important factors which contributed to a considerable extension of the domain of methodological studies, and caused even a fundamental change in the whole character of the methodology of deductive sciences. That conception of methodology which was indicated at the beginning of the present chapter has, during the historical development of the subject, turned out to be too narrow. The analysis and critical evaluation of methods applied in practice in the construction of deductive sciences ceased to be the exclusive or even the main task of methodology. The methodology of the deductive sciences became a general science of deductive sciences in an analogous sense as arithmetic is the science of numbers and geometry is the science of geometrical configurations. In contemporary methodology we investigate deductive theories as wholes as well as the sentences which constitute them; we consider the symbols and expressions of which these sentences are composed, properties and sets of ex-

[6] These exceedingly important achievements are due to the contemporary Austrian logician K. Gödel. His results concerning the decision problem were further extended by the contemporary American logician A. Church.

pressions and sentences, relations holding among them (such as the consequence relation) and even relations between expressions and the things which the expressions "talk about" (such as the relation of designation); we establish general laws concerning these concepts.

*Hence it follows that the terms which denote expressions occurring in deductive theories, properties of these expressions and relations among them belong, not to the domain of logic, but to the methodology of deductive sciences. This applies in particular to several of the terms introduced and employed in the previous chapters of this book, such as "*variable*", "*sentential function*", "*quantifier*", "*consequence*" and many others. In order to make clearer to ourselves the difference between logical and methodological terms, let us consider such a pair of words as "*or*" and "*disjunction*". The word "*or*" belongs, of course, to logic—namely, to sentential calculus—, although it is also used in all other sciences, and thus in particular in methodology. The word "*disjunction*", on the other hand, which denotes sentences constructed with the help of the word "*or*", is a typical instance of a methodological term.

The reader will perhaps be surprised at the fact that in the chapters concerned with logic we employed so many methodological terms. The explanation of this, however, is relatively simple. On the one hand, a certain circumstance here plays a role to which we have already called attention in Section 9: there is a widespread custom among logicians as well as mathematicians—sometimes for purely stylistic reasons—of using phrases which contain methodological terms as synonyms for expressions of a purely logical or mathematical character; we have to some extent complied with this custom in the present work. But, on the other hand, a more important factor is also involved here; we have not attempted in this book to construct logic in a systematic way, but have merely talked about logic and have discussed and commented on its concepts and laws. We know however (from Section 18) that when talking about logical concepts we must use the names of these concepts—thus the terms which already belong to methodology. Should we develop logic in the form of a deductive theory, without making any comments on it at all, then methodological terms would occur only in the formulation of rules of definition and inference.*

In connection with the evolution through which methodology has passed there has arisen a need for applying new, more subtle and more precise methods of inquiry in this field. Methodology has become like those sciences which constitute its own subject matter— it has assumed the form of a deductive discipline. In view of the extended domain of investigations, the expression "the methodology of deductive sciences" itself has ceased to appear appropriate enough; indeed, "methodology" means merely "the science of method". Consequently this expression is now often replaced by others—for instance, by a (not altogether happy) term "THEORY OF PROOF", or by a (much better) term "META-LOGIC AND META-MATHEMATICS", which means about the same as "the science of logic and mathematics". Still another term has been recently coming into use, "LOGICAL SYNTAX AND SEMANTICS OF DEDUCTIVE SCIENCES", which stresses the analogy between the methodology of deductive sciences and the grammar of everyday language.[7]

Exercises

1. The calculus of classes which was considered in Chapter IV can be constructed as a separate deductive theory, presupposing sentential calculus only. In this construction we shall consider the symbols "V", "Λ", "⊂" and all operation signs introduced in Section 25 as primitive terms. We assume, further, the following nine axioms[8]:

[7] Methodology of the deductive sciences in its widened meaning is a very young discipline. Its intensive development began only about twenty years ago—simultaneously (and, as it seems, independently) in two different centers: Göttingen under the influence of D. HILBERT (cf. footnote 1 on p. 120) and P. BERNAYS, and Warsaw where S. LEŚNIEWSKI and J. ŁUKASIEWICZ, among others, worked (cf. footnote 3 on p. 133 and footnote 2 on p. 19).

[8] The axiom system given here is essentially due to SCHRÖDER (cf. footnote 1 on p. 88). Various simple and interesting axiom systems for the calculus of classes were published by the contemporary American mathematician E. V. HUNTINGTON, to whom we are indebted for many important contributions concerning the axiomatic foundations of logical and mathematical theories.

Axiom I. $K \subset K$.

Axiom II. *If* $K \subset L$ *and* $L \subset M$, *then* $K \subset M$.

Axiom III. $K \cup L \subset M$ *if, and only if,* $K \subset M$ *and* $L \subset M$.

Axiom IV. $M \subset K \cap L$ *if, and only if,* $M \subset K$ *and* $M \subset L$.

Axiom V. $K \cap (L \cup M) \subset (K \cap L) \cup (K \cap M)$.

Axiom VI. $K \subset \vee$.

Axiom VII. $\wedge \subset K$.

Axiom VIII. $\vee \subset K \cup K'$.

Axiom IX. $K \cap K' \subset \wedge$.

From these axioms we may derive various theorems. Prove, in particular, the following theorems (making use of the hints which follow them):

Theorem I. $K \cup K \subset K$.

Hint: In Axiom III replace "L" and "M" by "K". Notice that the right side of the equivalence thus obtained is satisfied by any class K (Axiom I); therefore the left side must also be always satisfied.

Theorem II. $K \subset K \cap K$.

Hint: The proof based upon Axioms IV and I is analogous to that of Theorem I.

Theorem III. $K \subset K \cup L$ *and* $L \subset K \cup L$.

Hint: In Axiom III put "$K \cup L$" in place of "M"; notice that the left side of the equivalence, by Axiom I, is always satisfied.

Theorem IV. $K \cap L \subset K$ *and* $K \cap L \subset L$.

Hint: The proof is analogous to that of Theorem III.

Theorem V. $K \cup L \subset L \cup K$.

Hint: In Axiom III substitute "$L \cup K$" for "M"; and compare the right side of the equivalence thus obtained with Theorem III (where "K" is to be replaced by "L", and "L" by "K").

THEOREM VI. $K \cap L \subset L \cap K.$

Hint: The proof based upon Axiom IV and Theorem IV is analogous to that of Theorem V.

THEOREM VII. *If* $L \subset M$, *then* $K \cup L \subset K \cup M.$

Hint: Assuming that the hypothesis of the theorem is satisfied, derive the formulas:

$$K \subset K \cup M \quad \text{and} \quad L \subset K \cup M$$

(The first of these formulas follows directly from Theorem III, and the second can be deduced from the hypothesis and Theorem III by Axiom II). Apply Axiom III to these formulas.

THEOREM VIII. *If* $L \subset M$, *then* $K \cap L \subset K \cap M.$

Hint: The proof is similar to that of the preceding theorem.

THEOREM IX. $K \cap L \subset K \cap (L \cup M)$ *and* $K \cap M \subset K \cap (L \cup M).$

Hint: In Theorem III replace "K" by "L" and "L" by "M"; to the formulas thus obtained apply Theorem VIII.

THEOREM X. $(K \cap L) \cup (K \cap M) \subset K \cap (L \cup M).$

Hint: This theorem can be derived from Axiom III and Theorem IX.

Axioms III and IV, which play the most important role in the proofs of the above theorems, are called LAWS OF COMPOSITION (for addition and multiplication of classes).

2. Into the calculus of classes whose construction was outlined in the preceding exercise we may introduce the identity sign "=", defining it as follows:

DEFINITION I. $K = L$ *if, and only if,* $K \subset L$ *and* $L \subset K.$

From the axioms and theorems of Exercise 1 and the above definition derive the following theorems:

THEOREM XI. $K = K.$

Hint: Put "K" in place of "L" in Definition I and apply Axiom I.

THEOREM XII. *If* $K = L$, *then* $L = K$.

Hint: In Definition I replace "K" by "L" and "L" by "K"; compare the sentence thus obtained with Definition I in its original formulation.

THEOREM XIII. *If* $K = L$ *and* $L = M$, *then* $K = M$.

Hint: This theorem can be derived from Definition I and Axiom II.

THEOREM XIV. $K \cup K = K$.

Hint: In Definition I replace "K" by "$K \cup K$" and "L" by "K"; apply Theorem I and Theorem III (with "K" put in place of "L")

THEOREM XV. $K \cap K = K$.

Hint: The proof is analogous to that of the preceding theorem.

THEOREM XVI. $K \cup L = L \cup K$.

Hint: By Theorem V we have:

$$K \cup L \subset L \cup K \quad \text{and also} \quad L \cup K \subset K \cup L.$$

To these formulas apply Definition I.

THEOREM XVII. $K \cap L = L \cap K$.

Hint: The proof is similar to that of Theorem XVI.

THEOREM XVIII. $K \cap (L \cup M) = (K \cap L) \cup (K \cap M)$.

Hint: This theorem is a consequence of Definition I, Axiom V and Theorem X.

THEOREM XIX. $K \cup K' = \vee$.

Hint: This theorem can be derived, with the help of Definition I, from Axiom VI (with "K" replaced by "$K \cup K'$") and Axiom VIII.

THEOREM XX. $K \cap K' = \wedge$.

Hint: Apply Definition I, Axiom VII and Axiom IX.

Notice which of the axioms and theorems of this exercise and the preceding one are known to us from Chapter IV (or, possibly, III); recall their names.

3. Let us assume that, into the system of the calculus of classes discussed in Exercises 1 and 2 we introduce a new symbol "$\between$" denoting a certain relation between classes and defined as follows:

$$K \between L \textit{ if, and only if, neither } K \subset L \textit{ nor } L \subset K \textit{ nor } K \cap L = \wedge.$$

Is the relation defined in this way identical with any of the relations defined in Section 24?

Let the relation of disjointness between classes be denoted by the symbol ")(". How can this symbol be defined within our system of the calculus of classes?

4. Exhibit several interpretations within arithmetic and geometry of the axiom system considered in Section 37.

Is the set of all numbers, together with the relation *less than* among numbers, a model of this axiom system? Is the set of all straight lines and the relation of parallelism among lines such a model?

5. In that fragment of geometry which was discussed in Section 37, the relation of being shorter among segments can be defined in the following way:

We say that x is shorter than y, in symbols: $x < y$, *if x and y are segments and if x is congruent to a segment which is a part of y*; in other words, *if* $x \in \mathrm{S}$, $y \in \mathrm{S}$, *and if there exists a z such that* $z \in \mathrm{S}$, $z \subset y$, $z \neq y$ *and* $x \cong z$.

Differentiate in this sentence between the definiendum and the definiens; determine the disciplines (or the parts of logic, as the case may be) to which the terms occurring in the definiens belong. Does this definition comply with the general methodological principles of Section 36 and the rules of definition of Section 11?

6. Is the proof of Theorem I, as given in Section 37, a complete proof, if only those rules of proof are taken into consideration that were stated in Section 15?

7. In addition to Theorems I and II, the following theorems can be derived from the axioms of Section 37:

THEOREM III. *For any elements x, y and z of the set* **S**, *if* $x \cong y$ *and* $x \cong z$, *then* $y \cong z$.

THEOREM IV. *For any elements x, y and z of the set* **S**, *if* $x \cong y$ *and* $y \cong z$, *then* $z \cong x$.

THEOREM V. *For any elements x, y, z and t of the set* **S**, *if* $x \cong y$, $y \cong z$ *and* $z \cong t$, *then* $x \cong t$.

Give a strict proof that the following systems of sentences are equipollent, in the sense established in Section 39, to the system consisting of Axioms I and II (and that each might, therefore, be chosen as a new axiom system):

(a) the system consisting of Axiom I and Theorems I and II;

(b) the system consisting of Axiom I and Theorem III;

(c) the system consisting of Axiom I and Theorem IV;

(d) the system consisting of Axiom I and Theorems I and V.

8. Along the lines of the remarks made in Section 37 formulate general laws of the theory of relations that represent a generalization of the results obtained in the preceding exercise.

Hint: These laws may, for instance, be given the form of equivalences, beginning with the words:

for a relation R to be reflexive and to have the property **P** *in a class K it is necessary and sufficient that...*

9. Consider the system of sentences (a) of Exercise 7. Exhibit models satisfying

(a) the first two sentences of the system, but not the last;

(b) the first and third sentence, but not the second;

(c) the last two sentences, but not the first.

What conclusion may be drawn from the fact of the existence of such models with respect to the possibility of deriving any one of the three sentences from the others? Are these sentences mutually independent? (Cf. Sections 37 and 39.)

10. There have occasionally been complaints to the effect that there is a certain discrepancy in the various textbooks of geometry, inasmuch as sentences treated as theorems in some textbooks are adopted as axioms, and thus without proof, in others. Are these complaints justified?

*11. In Section 13 we became acquainted with the method of truth tables, which enables us in any particular case to decide whether a given sentence of sentential calculus is true and whether it may be, therefore, accepted as a law of this calculus. When applying this method, we may entirely forget about the meaning which was ascribed to the symbols "T" and "F" occurring in the truth tables; we may assume that this method reduces to applying, in the construction of sentential calculus, two rules, the first of which is close to the rules of definition and the second to the rules of proof. According to the first rule, if we want to introduce into sentential calculus a constant term, we must begin by constructing the fundamental truth table for the simplest—and at the same time the most general—sentential function containing this term. According to the second rule, if we want to accept a sentence (containing only those constants for which the fundamental truth tables have already been constructed) as a law of sentential calculus, we must construct the derivative truth table for this sentence and verify that the symbol "F" never occurs in the last column of this table.

Sentential calculus constructed exclusively by means of these two rules assumes a character close to that of formalized deductive theories. Justify this statement on the basis of the considerations of Section 40. Notice, however, some differences between this method of constructing sentential calculus and the general principles of constructing deductive theories, which were discussed in Section 36. With the method under consideration, is it possible to differentiate in sentential calculus between primitive and defined terms? What other distinction is lost here?

*12. Applying the method of truth tables as it was described in the preceding exercise, we may introduce into sentential calculus new terms which were not discussed in Chapter II. We can, for instance, introduce the symbol "Δ", considering the sentential function:

$$p \mathbin{\Delta} q$$

as an abbreviation of the expression:

neither p nor q.

Construct the fundamental truth table for this function, which would comply with the intuitive meaning ascribed to the symbol "Δ", and then verify, with the help of the derivative truth tables, that the following sentences are true and may be accepted as laws of sentential calculus:

$$(\sim p) \leftrightarrow (p \mathbin{\Delta} p),$$

$$(p \lor q) \leftrightarrow [(p \mathbin{\Delta} q) \mathbin{\Delta} (p \mathbin{\Delta} q)],$$

$$(p \rightarrow q) \leftrightarrow \{[(p \mathbin{\Delta} p) \mathbin{\Delta} q] \mathbin{\Delta} [(p \mathbin{\Delta} p) \mathbin{\Delta} q]\}.$$

*13. There exists a method of constructing sentential calculus as a formalized deductive theory, which differs from that described in Exercise 11, and which complies entirely with all the principles presented in Sections 36 and 40.[9] We may, for instance, assume the symbols "→", "↔" and "∼" (cf. Section 13) as primitive terms and the following seven sentences as axioms of sentential calculus:

Axiom I. $p \rightarrow (q \rightarrow p)$

Axiom II. $[p \rightarrow (p \rightarrow q)] \rightarrow (p \rightarrow q)$

Axiom III. $(p \rightarrow q) \rightarrow [(q \rightarrow r) \rightarrow (p \rightarrow r)]$

Axiom IV. $(p \leftrightarrow q) \rightarrow (p \rightarrow q)$

Axiom V. $(p \leftrightarrow q) \rightarrow (q \rightarrow p)$

Axiom VI. $(p \rightarrow q) \rightarrow [(q \rightarrow p) \rightarrow (p \leftrightarrow q)]$

Axiom VII. $[(\sim q) \rightarrow (\sim p)] \rightarrow (p \rightarrow q)$

Furthermore, we agree to apply in proofs two rules of inference with which we are already familiar, namely the rules of substitu-

[9] This method originates with Frege (cf. footnote 2 on p. 19).

tion and of detachment. (In order to formulate these rules, and especially the rule of substitution, quite exactly, we should have to establish the manner of using parentheses and to specify which expressions are to be considered as sentential functions in our calculus and may be therefore substituted for variables; this task presents no great difficulty.)

With the help of these rules of inference we are now in a position to deduce various theorems from our axioms. Give, in particular, complete proofs for the following theorems (making use of the hints which follow them):

THEOREM I. $p \rightarrow p$

Hint: Substitute "p" for "q" in Axioms I and II; notice that the first sentence thus obtained coincides with the antecedent of the second, and accordingly apply the rule of detachment.

THEOREM II. $p \rightarrow \{(p \rightarrow q) \rightarrow [(p \rightarrow q) \rightarrow q]\}$

Hint: In Axiom I substitute "$(p \rightarrow q)$" for "q"; in Axiom III replace "p", "q" and "r" by "$(p \rightarrow q)$", "p" and "q", respectively. Notice that the consequent of the first implication thus obtained coincides with the antecedent of the second. Now replace, in Axiom III, "q" by the antecedent of the second implication and "r" by its consequent (leaving "p", which is the antecedent of the first implication, unchanged). Then apply the rule of detachment twice.—This proof is a typical instance of reasoning based upon Axiom III, which is another form of the law of the hypothetical syllogism (cf. Section 12).

THEOREM III. $p \rightarrow [(p \rightarrow q) \rightarrow q]$

Hint: The proof is analogous to that of Theorem II. From Axiom II derive, by substitution, the sentence:

$$\{(p \rightarrow q) \rightarrow [(p \rightarrow q) \rightarrow q]\} \rightarrow [(p \rightarrow q) \rightarrow q].$$

Compare the antecedent of this sentence with the consequent of Theorem II; accordingly, perform a suitable substitution on Axiom III and apply the rule of detachment twice.

THEOREM IV. $[p \rightarrow (q \rightarrow r)] \rightarrow [q \rightarrow (p \rightarrow r)]$

Hint: From Axiom III derive, by substitution, the sentence:

(1) $$[p \rightarrow (q \rightarrow r)] \rightarrow \{[(q \rightarrow r) \rightarrow r] \rightarrow (p \rightarrow r)\}.$$

Furthermore, replace, in Axiom III, "p", "q" and "r" by "q", "$[(q \rightarrow r) \rightarrow r]$" and "$(p \rightarrow r)$", respectively. Notice that the antecedent of the implication thus derived can be obtained by substitution from Theorem III. Perform this substitution and, by applying the rule of detachment, derive:

(2) $\{[(q \rightarrow r) \rightarrow r] \rightarrow (p \rightarrow r)\} \rightarrow [q \rightarrow (p \rightarrow r)]$.

Notice now that the consequent of (1) is the same as the antecedent of (2); and, accordingly, proceed as in the proof of Theorem II (by applying Axiom III again).—Theorem IV is called LAW OF COMMUTATION.

THEOREM V. $(\sim p) \rightarrow (p \rightarrow q)$

Hint: From Axiom I derive by substitution:

$$(\sim p) \rightarrow [(\sim q) \rightarrow (\sim p)].$$

Notice that the consequent of this sentence coincides with the antecedent of one of the axioms; and proceed as in the proof of Theorem II.

THEOREM VI. $p \rightarrow [(\sim p) \rightarrow q]$

Hint: Perform a substitution on Theorem IV such that the antecedent of the resulting implication will be Theorem V, and then apply the rule of detachment.—We have here a typical instance of reasoning based upon the law of commutation.

THEOREM VII. $[\sim(\sim p)] \rightarrow (q \rightarrow p)$

Hint: The proof is analogous to that of Theorem II. From Theorem V and Axiom VII derive the sentences:

$[\sim(\sim p)] \rightarrow [(\sim p) \rightarrow (\sim q)]$ and $[(\sim p) \rightarrow (\sim q)] \rightarrow (q \rightarrow p)$.

Compare the antecedents and the consequents of these sentences.

THEOREM VIII. $[\sim(\sim p)] \rightarrow p$

Hint: Reasoning as in the proof of Theorem VI, derive first, from Theorems IV and VII, the sentence:

$$q \rightarrow \{[\sim(\sim p)] \rightarrow p\}.$$

In this sentence put any one of our axioms in place of "q", and apply the rule of detachment.

THEOREM IX. $p \rightarrow [\sim(\sim p)]$

Hint: Perform suitable substitutions on Axiom VII and Theorem VIII so as to be able to apply the rule of detachment.

THEOREM X. $[\sim(\sim p)] \leftrightarrow p$

Hint: This theorem can be obtained from Axiom VI and Theorems VIII and IX by performing a substitution on Axiom VI and by applying the rule of detachment twice.

*14. If we are to be able to introduce defined terms into the system of sentential calculus described in the preceding exercise, we have to assume a rule of definition. According to this rule (cf. Section 11), every definition has the form of an equivalence. The definiendum is an expression containing, besides sentential variables, only one constant, namely the term to be defined; no symbol may occur in this expression twice. The definiens is an arbitrary sentential function containing exactly the same variables as the definiendum, and containing no constants except primitive terms and terms previously defined. Thus we may, for instance, accept the following definitions of the symbols " $\vee$ " and " $\wedge$ ":

DEFINITION I. $(p \vee q) \leftrightarrow [(\sim p) \rightarrow q]$

DEFINITION II. $(p \wedge q) \leftrightarrow \{\sim [(\sim p) \vee (\sim q)]\}$

From the above definitions and the axioms and theorems of Exercise 13 deduce the following theorems with the help of the rules of substitution and detachment:

THEOREM XI. $[(\sim p) \rightarrow q] \rightarrow (p \vee q)$

Hint: In Axiom V substitute "$(p \vee q)$" for "p" and "$[(\sim p) \rightarrow q]$" for "q"; compare the sentence thus obtained with Definition I and apply the rule of detachment.

THEOREM XII. $p \vee (\sim p)$

Hint: This theorem can be derived from Theorems XI and I by two applications of the rule of substitution and one application of the rule of detachment.

THEOREM XIII. $p \rightarrow (p \vee q)$

Hint: The proof is based upon Axiom III and Theorems VI and XI, and is quite similar to that of Theorem II (the rule of substitution is applied to Axiom III only).

THEOREM XIV. $(p \wedge q) \rightarrow \{\sim [(\sim p) \vee (\sim q)]\}$

Hint: The proof, which is based upon Axiom IV and Definition II, is analogous to that of Theorem XI.

THEOREM XV. $\{\sim [\sim (p \wedge q)]\} \rightarrow \{\sim [(\sim p) \vee (\sim q)]\}$

Hint: The proof is based upon Axiom III and Theorems VIII and XIV, and is similar to that of Theorem II. In Theorem VIII replace "p" by "$(p \wedge q)$" and compare the consequent of the implication thus obtained with the antecedent of Theorem XIV.

THEOREM XVI. $[(\sim p) \vee (\sim q)] \rightarrow [\sim (p \wedge q)]$

Hint: Perform a substitution on Axiom VII such that the antecedent of the resulting implication will be Theorem XV.

THEOREM XVII. $(\sim p) \rightarrow [\sim (p \wedge q)]$

Hint: The proof is again analogous to that of Theorem II. In Theorem XIII substitute "$(\sim p)$" for "p" and "$(\sim q)$" for "q"; compare the resulting sentence with Theorem XVI.

THEOREM XVIII. $(p \wedge q) \rightarrow p$

Hint: Derive this theorem from Axiom VII and Theorem XVII.

Notice which of the axioms and theorems of this exercise and the preceding one are familiar to us from Chapter II, and recall their names.

*15. Formulate a definition of the symbol "Δ" (cf. Exercise 12), complying with the rule of definition stated in the preceding exercise; in the definiens there should occur two constants: "$\sim$" and "$\wedge$".

*16. Verify by the method of truth tables that all the axioms and definitions given in Exercises 13 and 14 (and also the definition proposed in Exercise 15) are true sentences. Try to conclude from this that all theorems which can be derived from the above axioms and definitions by applying the rules of substitution and detachment must also be found to be true sentences when tested by the method of truth tables.

(It is possible to show that also conversely every sentence of sentential calculus whose truth can be verified by the method of truth tables is either one of the axioms and definitions, or is derivable from them by means of our rules of inference—and, consequently, that the two methods of constructing sentential calculus, which were discussed in Exercises 11 and 13–14, are completely equivalent. But this task is vastly more difficult.)

*17. One of the methods of constructing sentential calculus discussed in Exercises 11 and 13–14 provides an immediate solution of the decision problem (cf. Section 41) for this calculus, and enables us to show very easily that sentential calculus is a consistent deductive theory. Which is this method, and how can this be shown?

*18. One of the laws of sentential calculus is the following:

For any p and q, if p and not p, then q.

On the basis of this logical law, establish the following methodological law:

If the axiom system of any deductive theory which presupposes sentential calculus is inconsistent, then every sentence formulated in the terms of this theory can be derived from that system.

*19. It is known that the following methodological law holds:

If the axiom system of a deductive theory is complete, and if any sentence which can be formulated but not proved within that theory is added to the system, then the axiom system extended in this manner is no longer consistent.

Why is this the case?

*20. Single out all those terms appearing in Chapter II, which belong to the field of the methodology of deductive sciences, according to the remarks made in Section 42.

SECOND PART

APPLICATIONS OF LOGIC AND METHODOLOGY IN CONSTRUCTING MATHEMATICAL THEORIES

· VII ·

CONSTRUCTION OF A MATHEMATICAL THEORY:

LAWS OF ORDER FOR NUMBERS

43. Primitive terms of the theory under construction; axioms concerning fundamental relations among numbers

With a certain amount of knowledge of the fields of logic and methodology at our disposal, we shall now undertake to lay the foundations of a particular and, incidentally, very elementary mathematical theory. This will be a good opportunity for us to assimilate better our previously acquired knowledge, and even to expand it to some extent.

The theory with which we shall concern ourselves constitutes a fragment of the arithmetic of real numbers. It contains fundamental theorems concerning the basic relations *less than* and *greater than* among numbers, as well as the basic operations on numbers, namely of addition and subtraction. It presupposes nothing but logic.

The primitive terms which we shall adopt in this theory are the following:

real number,

is less than,

is greater than,

sum.

Instead of "*real number*" we shall, as before, simply say "*number*". Also, it is slightly more convenient to consider, instead of the term "*number*", the expression "*the set of all numbers*" as a primitive term, which, for brevity, we will replace by the symbol "**N**"; thus, in order to express that x is a number, we write:

$$x \in \mathbf{N}.$$

We may, on the other hand, stipulate that the universe of discourse of our theory consists of real numbers only and that variables such as "x", "y", ... stand exclusively for the names of numbers; in this case, the term "*real number*" would be altogether dispensable in the formulations of statements of our theory, and the symbol "**N**" might, when needed, be replaced by "**V**" (cf. Section 23).

The expressions "*is less than*" and "*is greater than*" are to be treated as if they were entities consisting of a single word each; they will be replaced by the briefer symbols "$<$" and "$>$", respectively. Instead of "*is not less than*" and "*is not greater than*" we shall employ the usual symbols "$\nless$" and "$\ngtr$". Further, instead of "*the sum of the numbers* (*the summands*) x *and* y" or "*the result of adding* x *and* y" we shall use the customary notation:

$$x + y.$$

Thus, the symbol "**N**" designates a certain set, the symbols "$<$" and "$>$" certain two-termed relations, and finally the symbol "$+$" a certain binary operation.

Among the axioms of the theory under consideration two groups may be distinguished. The axioms of the first group express fundamental properties of the relations *less than* and *greater than*, whereas those of the second are primarily concerned with addition. For the time being we shall consider the first group only; it consists, altogether, of five statements:

Axiom 1. *For any numbers* x *and* y (i.e., *for arbitrary elements of the set* **N**) *we have*: $x = y$ *or* $x < y$ *or* $x > y$.

Axiom 2. *If* $x < y$, *then* $y \nless x$.

Axiom 3. *If* $x > y$, *then* $y \ngtr x$.

Axiom 4. *If* $x < y$ *and* $y < z$, *then* $x < z$.

Axiom 5. *If* $x > y$ *and* $y > z$, *then* $x > z$.

The axioms listed here, just as any arithmetical theorem of a universal character stating that arbitrary numbers $x, y, \ldots$ have such and such a property, should really begin with the words "*for any numbers* $x, y, \ldots$" or "*for any elements* $x, y, \ldots$ *of the set* **N**" or, simply, "*for any* $x, y, \ldots$" (if we agree that the variables "x", "y", ... here denote numbers only). But since we want to conform to the usage discussed in Section 3, we often omit such a phrase and merely add it in our mind; this holds not only for the axioms but also for the theorems and definitions which will occur in the course of our considerations. Axiom 2, for instance, is meant to be read as follows:

For any x and y (or *for any elements x and y of the set* **N**), *if* $x < y$, *then* $y \nless x$.

We shall refer to Axiom 1 as the WEAK LAW OF TRICHOTOMY (with the strong law of trichotomy we shall become acquainted later). Axioms 2–5 express the fact that the relations *less than* and *greater than* are asymmetrical and transitive (cf. Section 29); accordingly they are called the LAWS OF ASYMMETRY and LAWS OF TRANSITIVITY for the relations *less than* and *greater than*. The axioms of the first group and the theorems following from them are called the LAWS OF ORDER FOR NUMBERS.

The relations $<$ and $>$, together with the logical relation of identity $=$, will be here referred to as the FUNDAMENTAL RELATIONS AMONG NUMBERS.

44. Laws of irreflexivity for the fundamental relations; indirect proofs

Our next task consists in the derivation of a number of theorems from the axioms adopted by us. Since we do not aim at a systematic presentation, in this and the following chapter only those theorems will be stated which may serve to help illustrate certain concepts and facts of the fields of logic and methodology.

THEOREM 1. *No number is smaller than itself*: $x \nless x$.

PROOF. Suppose our theorem were false. Then there would be a number x satisfying the formula:

(1) $$x < x$$

Now Axiom 2 refers to arbitrary numbers x and y (which need not be distinct), so that it remains valid if in place of "y" we write the variable "x"; we then obtain:

(2) *if* $x < x$, *then* $x \nless x$.

But from (1) and (2) it follows immediately that

$$x \nless x;$$

this consequence, however, forms an obvious contradiction to formula (1). We must, therefore, reject the original assumption and accept the theorem as proved.

We shall now show how to transform this argument into a complete proof, using for clarity the logical symbolism (cf. Sections 13 and 15). To this end we resort to the so-called LAW OF REDUCTIO AD ABSURDUM of sentential calculus:

(I) $[p \rightarrow (\sim p)] \rightarrow (\sim p)$[1]

We further use Axiom 2 in the following symbolic form:

(II) $(x < y) \rightarrow [\sim (y < x)]$

Our proof is based exclusively upon the sentences (I) and (II). First we apply the rule of substitution to (I), replacing "p" in it throughout by "$(x < x)$":

(III) $\{(x < x) \rightarrow [\sim (x < x)]\} \rightarrow [\sim (x < x)]$

We next apply the rule of substitution to (II), replacing "y" by "x":

(IV) $(x < x) \rightarrow [\sim (x < x)]$

Finally we observe that the sentence (IV) is the hypothesis of the conditional sentence (III), so that the rule of detachment may be applied. We are thus led to the formula:

(V) $\sim (x < x)$

which is the symbolic form of the theorem to be proved.

[1] This law, together with a related one of the same name:

$$[(\sim p) \rightarrow p] \rightarrow p,$$

has been used in many intricate and historically important arguments in logic and mathematics. The Italian logician and mathematician G. VAILATI (1863–1909) devoted a special monograph to its history.

The proof of Theorem 1 represents an example of what is called an INDIRECT PROOF, also known as a PROOF BY REDUCTIO AD ABSURDUM. Proofs of this kind may quite generally be characterized as follows: in order to prove a theorem, we assume the theorem to be false, and derive from that certain consequences which compel us to reject the original assumption. Indirect proofs are very common in mathematics. They do not all fall under the schema of the proof of Theorem 1, however; on the contrary, the latter represents a comparatively rare form of indirect proof, and we shall meet with more typical examples of indirect proofs further below.

The axiom system adopted by us is perfectly symmetrical with respect to the two symbols "$<$" and "$>$". To every theorem concerning the relation *less than*, we therefore automatically obtain the corresponding theorem concerning the relation *greater than*, the proofs being entirely analogous, so that the proof of the second theorem may be omitted altogether. In particular, corresponding to Theorem 1 we have:

THEOREM 2. *No number is greater than itself*: $x \ngtr x$.

While the relation of identity $=$, as we know from logic, is reflexive, Theorems 1 and 2 show that the other two fundamental relations among numbers, $<$ and $>$, are irreflexive; these theorems are therefore called the LAWS OF IRREFLEXIVITY (for the relations *less than* and *greater than*).

45. Further theorems on the fundamental relations

We shall next prove the following theorem:

THEOREM 3. $x > y$ *if, and only if,* $y < x$.

PROOF. It has to be shown that the formulas:

$$x > y \quad \text{and} \quad y < x$$

are equivalent, that is to say, that the first implies the second, and vice versa (cf. Section 10).

Suppose, first, that

(1) $$y < x.$$

By Axiom 1 we must have at least one of the three cases:

(2) $$x = y, \quad x < y \quad or \quad x > y.$$

If we had $x = y$, we could, by virtue of the fundamental law of the theory of identity, i.e. LEIBNIZ's law (cf. Section 17), replace the variable "x" by "y" in formula (1); the resulting formula:

$$y < y$$

constitutes an obvious contradiction to Theorem 1. Hence we have:

(3) $$x \neq y.$$

But we also have:

(4) $$x \not< y$$

since, by Axiom 2, the formulas:

$$x < y \quad \text{and} \quad y < x$$

cannot hold simultaneously. On account of (2), (3) and (4), we find that the third case must apply:

(5) $$x > y.$$

We thus have shown that the formula (5) is implied by the formula (1); conversely, the implication in the opposite direction can be established by an analogous procedure. The two formulas are, therefore, indeed equivalent, q.e.d.[2]

Using the terminology of the calculus of relations (cf. Section 28), we may say that, according to Theorem 3, each of the relations $<$ and $>$ is the converse of the other.

THEOREM 4. *If* $x \neq y$, *then* $x < y$ *or* $y < x$.

PROOF. Since

$$x \neq y,$$

we have, by Axiom 1:

$$x < y \quad or \quad x > y;$$

the second of these formulas implies, by Theorem 3:

$$y < x.$$

[2] The letters "q.e.d." are the customary abbreviation of the expression "quod erat demonstrandum", meaning "which was to be proved".

Hence we have:

$$x < y \quad \textit{or} \quad y < x, \qquad \text{q.e.d.}$$

Analogously we can prove

THEOREM 5. *If* $x \neq y$, *then* $x > y$ *or* $y > x$.

By Theorems 4 and 5 the relations $<$ and $>$ are connected; accordingly these theorems are known as the LAWS OF CONNEXITY (for the relations *less than* and *greater than*). Axioms 2–5, together with Theorems 4 and 5, show that the set of numbers **N** is ordered by either of the relations $<$ and $>$.

THEOREM 6. *Any numbers x and y satisfy one, and only one, of the three formulas:* $x = y$, $x < y$ *and* $x > y$.

PROOF. It follows from Axiom 1 that at least one of the formulas stated must be satisfied. In order to prove that the formulas:

$$x = y \quad \text{and} \quad x > y$$

exclude each other, we proceed as in the proof of Theorem 3: we replace in the second of these formulas "x" by "y" and arrive at a contradiction to Theorem 1. Similarly it can be shown that the formulas:

$$x = y \quad \text{and} \quad x > y$$

exclude each other. And finally, the two formulas:

$$x < y \quad \text{and} \quad x > y$$

cannot hold simultaneously, because, by Theorem 3, we would then have:

$$x < y \quad \text{and} \quad y < x,$$

in contradiction to Axiom 2. Hence, any numbers x and y satisfy one and no more of the three formulas in question, q.e.d.

Theorem 6 we will call the STRONG LAW OF TRICHOTOMY, or simply the LAW OF TRICHOTOMY; according to this law, one and only one of the three fundamental relations holds between any two given numbers. Using the phrase "*either ... or ...*" in the meaning proposed in Section 7, we can formulate Theorem 6 in a more concise manner:

For any numbers x and y we have either $x = y$ *or* $x < y$ *or* $x > y$.

46. Other relations among numbers

Apart from the fundamental relations, three other relations play an important part in arithmetic. One of these is the logical relation of diversity $\neq$ which we know already; the other are the relations $\leqq$ and $\geqq$ which will be discussed now.

The meaning of the symbol "$\leqq$" is explained by the following definition:

DEFINITION 1. *We say that* $x \leqq y$ *if, and only if,* $x = y$ *or* $x < y$.

The formula:

$$x \leqq y$$

is to be read: "*x is less than or equal to y*" or "*x is at most equal to y*".

Although the content of the definition as stated appears to be clear, experience shows that in practical applications it sometimes becomes the source of certain misunderstandings. Some people who believe they understand the meaning of the symbol "$\leqq$" perfectly well protest nevertheless against its application to definite numbers. They do not only reject a formula like:

$$1 \leqq 0$$

as obviously false—and this rightly so—, but they also consider as meaningless or even false such formulas as:

$$0 \leqq 0 \quad \text{or} \quad 0 \leqq 1;$$

for they maintain that there is no sense in saying that $0 \leqq 0$ or that $0 \leqq 1$ since it is known that $0 = 0$ and $0 < 1$. In other words, it is not possible to exhibit a single pair of numbers which, in their opinion, satisfies the formula:

$$x \leqq y.$$

This view is palpably mistaken. Just because $0 < 1$ holds, it follows that the sentence:

$$0 = 1 \quad or \quad 0 < 1$$

is true, for the disjunction of two sentences is certainly true provided one of them is true (cf. Section 7); but according to Definition 1 this disjunction is equivalent to the formula:

$$0 \leqq 1.$$

For a quite analogous reason the formula:

$$0 \leqq 0$$

is also true.

The source of these misunderstandings, presumably, lies in certain habits of everyday life (*to which we have already called attention at the end of Section 7*). In ordinary language it is customary to assert the disjunction of two sentences only if we know that one of the sentences is true without knowing which. It does not occur to us to say that $0 = 1$ *or* $0 < 1$, though this is undoubtedly true, since we can say something that is simpler and at the same time logically stronger, namely, that $0 < 1$. In mathematical considerations, however, it is not always advantageous to state everything that we know in its strongest possible form. For example, we sometimes assert of a quadrangle merely that it is a parallelogram, although we know it to be a square, and this because we may want to apply a general theorem concerning arbitrary parallelograms. For similar reasons it may occur that it is known of a number x (for instance, of the number 0) that it is less than 1, and yet it may merely be asserted that $x \leqq 1$, that is, that either $x = 1$ or $x < 1$.

We will now state two theorems concerning the relation $\leqq$.

THEOREM 7. $x \leqq y$ *if, and only if,* $x \not> y$.

PROOF. This theorem is an immediate consequence of Theorem 6, i.e. the law of trichotomy. In fact, if

(1) $$x \leqq y$$

and hence, by Definition 1,

(2) $$x = y \quad or \quad x < y,$$

it is impossible for the formula:

$$x > y$$

to hold. Conversely, if

(3) $$x \ngtr y,$$

we must have (2) and hence, again by Definition 1, formula (1) must hold. The formulas (1) and (3) are thus equivalent, q.e.d.

In the terminology of Section 28, Theorem 7 states that the relation $\leqq$ is the negation of the relation $>$.

On account of its structure, Theorem 7 might be looked upon as the definition of the symbol "$\leqq$"; it would be a different one from that adopted here but equivalent to it. The statement of this theorem may also contribute to dispel any last doubts about the usage of the symbol "$\leqq$"; for nobody will hesitate any longer to recognize as true such formulas as:

$$0 \leqq 0 \quad \text{and} \quad 0 \leqq 1$$

in view of the fact that they are equivalent to the formulas:

$$0 \ngtr 0 \quad \text{and} \quad 0 \ngtr 1.$$

If we wished, we could avoid the use of the symbol "$\leqq$" completely, by always employing "$\ngtr$" instead.

THEOREM 8. $x < y$ *if, and only if,* $x \leqq y$ *and* $x \neq y$.

PROOF. If

(1) $$x < y,$$

then, by Definition 1,

(2) $$x \leqq y$$

while, by the law of trichotomy, the formula:

$$x = y$$

cannot hold. Conversely, if formula (2) holds, then by Definition 1 we obtain:

(3) $$x < y \quad \textit{or} \quad x = y;$$

but if, at the same time, we have:

$$x \neq y,$$

we have to accept the first part of the disjunction (3), that is, formula (1). The implication therefore holds in both directions, q.e.d.

A number of other theorems concerning the relation $\leqq$ we shall pass over; among them, there are, in particular, theorems to the effect that this relation is reflexive and transitive. The proofs of none of these theorems afford any difficulties.

The definition of the symbol "$\geqq$" is entirely analogous to Definition 1; and from the theorems concerning the relation $\leqq$ we automatically obtain corresponding theorems concerning the relation $\geqq$ by merely replacing the symbols "$\leqq$", "$<$" and "$>$" throughout by the symbols "$\geqq$", "$>$" and "$<$".

Formulas of the form:

$$x = y$$

in which the places of "x" and "y" may be taken by constants, variables or compound expressions denoting numbers are usually called EQUATIONS. Similar formulas of the form:

$$x < y \quad \text{or} \quad x > y$$

are called INEQUALITIES (IN THE NARROWER SENSE); among the INEQUALITIES IN THE WIDER SENSE we have, in addition, formulas of the form:

$$x \neq y, \qquad x \leqq y \quad \text{or} \quad x \geqq y.$$

The expressions occurring on the left and right sides of the symbols "$=$", "$<$", and so on, in these formulas are referred to as the LEFT AND RIGHT SIDES OF THE EQUATION OR OF THE INEQUALITY.

Exercises

1. Consider two relations among men: that of being of a smaller stature, and that of being of a larger stature. What condition has to be satisfied by an arbitrary set of people, so that it together with those two relations forms a model of the first group of axioms (cf. Section 37)?

2. Let the formula:

$$x \circledS y$$

express the fact that the numbers x and y satisfy one of the following conditions: (i) the number x has a smaller absolute value than the number y, or (ii) if the absolute values of x and y are the same, x is negative and y is positive. Further, let the formula:

$$x \mathbin{⧁} y$$

have the same meaning as the formula:

$$y \mathbin{⧀} x.$$

Show, on the basis of arithmetic, that the set of all numbers and the relations ⧀ and ⧁ just defined constitute a model of the first group of axioms.

Give other examples of interpretations of these axioms within arithmetic and geometry.

3. From Theorem 1 derive the following theorem:

$$\textit{if} \quad x < y, \quad \textit{then} \quad x \neq y.$$

Conversely, derive Theorem 1 from the theorem just stated, without making use of any other arithmetical statements. Are these two inferences indirect and do they fall under the schema of the proof of Theorem 1 of Section 44?

4. Generalize the proof of Theorem 1 of Section 44, and thereby establish the following general law of the theory of relations (cf. remarks made in Section 37):

every relation R which is asymmetrical in the class K is also irreflexive in that class.

5. Show that, if Theorem 1 is adopted as a new axiom, the old Axiom 2 can be derived as a theorem from this axiom together with Axiom 4.

As a generalization of this argument, prove the following general law of the theory of relations:

every relation R which is irreflexive and transitive in the class K is also asymmetrical in that class.

*6. At the end of Section 44 we tried to explain why the proof of Theorem 2 may be omitted. These remarks represent an application of certain general considerations of Chapter VI. Explain this in detail, and, in particular, specify the considerations to which this refers.

7. Derive the following theorems from the first group of axioms:

(a) $x = y$ *if, and only if,* $x \nless y$ *and* $y \nless x$;

(b) *if* $x < y$, *then* $x < z$ *or* $z < y$.

8. Derive the following theorems from Axiom 4 and Definition 1:

(a) *if* $x < y$ *and* $y \leqq z$, *then* $x < z$;

(b) *if* $x \leqq y$ *and* $y < z$, *then* $x < z$;

(c) *if* $x \leqq y$, $y < z$ *and* $z \leqq t$, *then* $x < t$.

9. Show that the relations $\leqq$ and $\geqq$ are reflexive, transitive and connected. Are these relations symmetrical or asymmetrical?

10. Show that, between any two numbers, exactly three of the following six relations hold: $=$, $<$, $>$, $\neq$, $\leqq$ and $\geqq$.

11. Both the converse and the negation of any of the relations listed in the preceding exercise are again among these six relations. Show in detail that this is the case.

*12. Between which of the relations given in Exercise 10 does the relation of inclusion hold? What will be the sum, the product and the relative product of any pair among these relations?

Hint: Recall the terms explained in Section 28. Do not omit to consider pairs consisting of two equal relations, and remember that the relative product may depend upon the order of the factors (cf. Exercise 5 of Chapter V). Altogether 36 pairs of relations should be examined.

· VIII ·

CONSTRUCTION OF A MATHEMATICAL THEORY:

LAWS OF ADDITION AND SUBTRACTION

47. Axioms concerning addition; general properties of operations, concepts of a group and of an Abelian group

We now turn to the second group of axioms, which consists of the following six sentences:

AXIOM 6. *For any numbers y and z there exists a number x such that* $x = y + z$; in other words: *if* $y \in \mathbf{N}$ *and* $z \in \mathbf{N}$, *then also* $y + z \in \mathbf{N}$.

AXIOM 7. $x + y = y + x$.

AXIOM 8. $x + (y + z) = (x + y) + z$.

AXIOM 9. *For any numbers x and y there exists a number z such that* $x = y + z$.

AXIOM 10. *If* $y < z$, *then* $x + y < x + z$.

AXIOM 11. *If* $y > z$, *then* $x + y > x + z$.

For the moment let us concentrate on the first four sentences of this second group, that is, Axioms 6–9. They ascribe to the operation of addition a number of simple properties which are also frequently met when considering other operations in various parts of logic and mathematics.

Special terms have been introduced to designate these properties. Thus we say that the operation O is PERFORMABLE IN THE CLASS K or that the class K is CLOSED UNDER THE OPERATION O, if the performance of the operation O on any two elements of the class K results again in an element of that same class; in other words, if, for any two elements y and z of the class K, there exists an element x of this class such that

$$x = y\,O\,z.$$

The operation O is called COMMUTATIVE IN THE CLASS K, if the result of this operation is independent of the order of the elements of the class K on which it is carried out, or, in other words, if for any two elements x and y of the class we have:

$$x\,O\,y = y\,O\,x.$$

The operation O is ASSOCIATIVE IN THE CLASS K, if the result is independent of the way in which the elements are grouped together, or, more precisely, if for any three elements x, y and z of the class the condition:

$$x\,O\,(y\,O\,z) = (x\,O\,y)\,O\,z$$

is satisfied. The operation O is said to be RIGHT-INVERTIBLE or LEFT-INVERTIBLE IN THE CLASS K, if, for any two elements x and y of the class K, there always exists an element z of the class such that

$$x = y\,O\,z \quad \text{or} \quad x = z\,O\,y,$$

respectively, holds. An operation O which is both right- and left-invertible is simply called INVERTIBLE IN THE CLASS K. It follows at once that a commutative operation which is right- or left-invertible must be invertible. We shall now say that a class K is a GROUP WITH RESPECT TO THE OPERATION O, if this operation is performable, associative and invertible in K; if, moreover, the operation O is commutative, the class K is called an ABELIAN GROUP WITH RESPECT TO THE OPERATION O. The concept of a group and, in particular, that of an Abelian group, forms the subject of a special mathematical discipline known as the THEORY OF GROUPS, which has already been mentioned above in Chapter V.[1]

[1] The group concept was introduced into mathematics by the French mathematician E. GALOIS (1811–1832). The term "Abelian group" was chosen in honor of the Norwegian mathematician N. H. ABEL (1802–1829) whose researches have had a great influence upon the development of higher algebra. The far-reaching importance of the group concept for mathematics has been recognized particularly since the works of another Norwegian mathematician, S. LIE (1842–1899).

In case the class K is the universal class (or the universe of discourse of the theory considered—cf. Section 23), we usually omit the reference to this class when employing such terms as "performable", "commutative", and so on.

In accordance with the terminology introduced above, the Axioms 6–9 are referred to as the LAW OF PERFORMABILITY, the COMMUTATIVE LAW, the ASSOCIATIVE LAW and the LAW OF RIGHT INVERTIBILITY for the operation of addition, respectively; together they state that the set of all numbers constitutes an Abelian group with respect to addition.

48. Commutative and associative laws for a larger number of summands

Axiom 7, the commutative law, and Axiom 8, the associative law, in the form in which they have been stated here, refer to two and three numbers, respectively. But there are infinitely many other commutative and associative laws concerning more than two or three numbers. The formula:

$$x + (y + z) = y + (z + x),$$

for instance, constitutes an example of a commutative law for three summands, and the formula:

$$x + [y + (z + u)] = [(x + y) + z] + u$$

represents one of the associative laws for four summands. In addition, there are theorems of a mixed character which, generally expressed, assert that any changes in either the order or the grouping of the summands are without influence upon the result of the addition. By way of an example the following theorem may be stated.

THEOREM 9. $x + (y + z) = (x + z) + y.$

PROOF. By suitable substitutions we obtain from Axioms 7 and 8:

$$(1) \qquad z + y = y + z,$$

$$(2) \qquad x + (z + y) = (x + z) + y.$$

In view of (1) we may, in accordance with LEIBNIZ's law, replace "$z + y$" in (2) by "$y + z$"; the result is the desired formula:

$$x + (y + z) = (x + z) + y.$$

In a similar manner we can derive all commutative and associative laws concerning an arbitrary number of summands from Axioms 7 and 8 together, possibly, with Axiom 6. These theorems are often used in practice in the transformation of algebraic expressions. By a transformation of an expression denoting a number we mean, as usual, an alteration of such a kind as to lead to an expression denoting the same number, which may hence be joined with the original expression by the identity sign; the expressions most frequently subjected to transformations of this kind are those which contain variables and which, therefore, are designatory functions. On the basis of the commutative and associative laws we are in a position to transform any expressions of a form such as:

$$x + (y + z), x + [y + (z + u)], \dots,$$

that is, expressions consisting of numerical constants and variables separated by addition signs and parentheses; in any such expression we may interchange at will both the numerical symbols and the parentheses (provided only the resulting expression has not become meaningless on account of the transposition of the parentheses).

49. Laws of monotony for addition and their converses

Axioms 10 and 11, to which we will turn now, are the so-called LAWS OF MONOTONY for addition with respect to the relations *less than* and *greater than*. We say, more generally, that the binary operation O is MONOTONIC IN THE CLASS K WITH RESPECT TO THE TWO-TERMED RELATION R, if, for any elements x, y, z of the class K, the formula:

$$y\, R\, z$$

implies:

$$(x\, O\, y)\, R\, (x\, O\, z),$$

which means that the result of performing the operation O on x and y has the relation R to the result of performing the operation O on x and z. (In the case of non-commutative operations one should, strictly speaking, differentiate between right and left monotony, the one just defined being denoted as right monotony.)

The operation of addition is monotonic not only with respect to the relations *less than* and *greater than*—a consequence of Axioms 10 and 11—but also with respect to the other relations among numbers discussed in Section 46. We shall show this here only for the relation of identity:

THEOREM 10. *If* $y = z$, *then* $x + y = x + z$.

PROOF. The sum $x + y$, whose existence is guaranteed by Axiom 6, is equal to itself (by Law II of Section 17):

$$x + y = x + y.$$

In view of the hypothesis of the theorem, the variable "y" on the right side of this equation may be replaced by the variable "z", and we obtain the desired formula:

$$x + y = x + z.$$

The converse of Theorem 10 is also true:

THEOREM 11. *If* $x + y = x + z$, *then* $y = z$.

We shall sketch two proofs of this theorem here. The first, based upon the law of trichotomy and Axioms 6, 10 and 11, is comparatively simple. For our later aims we require, however, another proof which is considerably more involved, but does not make use of anything except Axioms 7–9.

FIRST PROOF. Suppose the theorem in question were false. Then there would be numbers x, y and z such that

$$(1) \qquad x + y = x + z$$

and yet

$$(2) \qquad y \neq z.$$

Since $x + y$ and $x + z$ are numbers (according to Axiom 6), they can, by the law of trichotomy, satisfy only one of the formulas:

$$x + y = x + z, \quad x + y < x + z \quad \text{and} \quad x + y > x + z.$$

Since, by (1), the first holds, the others are automatically eliminated. We therefore have:

$$(3) \qquad x + y \nless x + z \quad and \quad x + y \ngtr x + z.$$

By applying the law of trichotomy once more, we can, on the other hand, infer from the inequality (2) that

$$y < z \quad or \quad y > z.$$

Hence, by Axioms 10 and 11,

$$(4) \qquad x + y < x + z \quad or \quad x + y > x + z,$$

which represents an obvious contradiction to (3). The supposition is thus refuted, and the theorem must be considered proved.

*SECOND PROOF. Apply Axiom 9, with "x" and "z" replaced by "y" and "u", respectively. It follows that there exists a number u fulfilling the formula:

$$y = y + u.$$

Since, by Axiom 7,

$$y + u = u + y,$$

we have, on account of the transitivity of the relation of identity (cf. Law IV of Section 17):

$$(1) \qquad y = u + y.$$

Now apply Axiom 9 again, with "x" and "z" replaced by "z" and "v", respectively; we thereby obtain a number v satisfying the equation:

$$(2) \qquad z = y + v.$$

On account of (1), we may here replace the variable "y" by the expression "$u + y$":

$$z = (u + y) + v.$$

Further, by the associative law, i.e. Axiom 8, we have:

$$u + (y + v) = (u + y) + v,$$

so that, by applying Law V of Section 17, we arrive at:

$$z = u + (y + v).$$

On account of (2), we may here replace "$y + v$" by "z" (using LEIBNIZ's law), so that we finally obtain:

$$(3) \qquad z = u + z.$$

Applying Axiom 9 for the third time, this time with "x", "y" and "z" replaced by "u", "x" and "w", respectively, we obtain a number w for which

$$u = x + w$$

holds, and since

$$x + w = w + x,$$

we have:

$$(4) \qquad u = w + x.$$

Using (4) we obtain the following formula from (1):

$$y = (w + x) + y;$$

but since, by the associative law, we have:

$$w + (x + y) = (w + x) + y,$$

this formula becomes:

$$(5) \qquad y = w + (x + y).$$

In view of the hypothesis of the theorem to be proved, we may replace "$x + y$" in (5) by "$x + z$", which leads to:

$$(6) \qquad y = w + (x + z).$$

Applying again the associative law, we have:

$$w + (x + z) = (w + x) + z,$$

so that (6) becomes:

$$y = (w + x) + z.$$

On account of (4), we may here replace "$w + x$" by "u". In this way we obtain:

$$(7) \qquad y = u + z.$$

But from equations (7) and (3) it follows that

$$y = z,$$ q.e.d.*

A few remarks concerning the first proof of Theorem 11 may be inserted here. Like the proof of Theorem 1, it constitutes an example of an indirect inference. The schema of this proof may be represented as follows. In order to prove a certain sentence, say "*p*", we suppose the sentence to be false, that is we assume the sentence "*not p*". From this assumption a consequence "*q*" is derived; that is to say, we demonstrate the implication:

if not p, then q

(in the case under consideration, the consequence "*q*" is the conjunction of the conditions (3) and (4) which appear in the proof). On the other hand, however, we are able to show (either on the basis of general laws of logic, as in the case under consideration, or by some theorems previously proved within the mathematical discipline in which all these arguments are carried out), that the consequence obtained is false, that is, that "*not q*" holds; thereby we are compelled to give up the original assumption, and thus to accept the sentence "*p*" as true. If this argument were set down in the form of a complete proof, a logical law which would play an essential part in it is a variant of the law of contraposition known from Section 14, and which reads as follows:

From: if not p, then q, it follows that: if not q, then p.

The proof under consideration differs slightly from that of Theorem 1. There, from the assumption that the theorem is false, we inferred that the theorem is true, that is, we derived a consequence directly contradicting the assumption; here, however, we derived from a similar assumption a consequence of which we knew from other sources that it was false. But this difference is not an essential one; it can easily be seen on the basis of logical laws that the proof of Theorem 1—like any other indirect mode of inference—can be brought under the schema sketched above.

Like Theorem 10, the other laws of monotony, that is, Axioms 10 and 11, also admit of conversion:

Theorem 12. *If* $x + y < x + z$, *then* $y < z$.

Theorem 13. *If* $x + y > x + z$, *then* $y > z$.

The proof of these theorems can without difficulty be obtained along the lines of the proof of Theorem 1.

50. Closed systems of sentences

There exists a general logical law the knowledge of which considerably simplifies the proofs of the last three theorems (11, 12 and 13). This law, sometimes called the LAW OF CLOSED SYSTEMS or HAUBER'S LAW[2], permits us in some cases, when we have succeeded in proving several conditional sentences, to infer from the form of these sentences that the corresponding converse sentences may be also considered as proved.

Suppose we are given a number of implications, say three, to which we will give the following schematic form:

if p_1, *then* q_1;
if p_2, *then* q_2;
if p_3, *then* q_3.

These three sentences are said to form a CLOSED SYSTEM, if their antecedents are of such a kind as to exhaust all possible cases, that is, if is true that:

p_1 *or* p_2 *or* p_3,

and if, at the same time, their consequents exclude one another:

if q_1, *then not* q_2; *if* q_1, *then not* q_3; *if* q_2, *then not* q_3.

The law of closed systems asserts that if certain conditional sentences forming a closed system are true, then the corresponding converse sentences are also true.

The simplest example of a closed system is given in the form of a system of two sentences, consisting of some implication:

if p, then q,

and its inverse sentence:

if not p, then not q.

[2] After the name of the German mathematician K. F. HAUBER (1775–1851).

In order to demonstrate the two converse sentences in this case, it is not even necessary to resort to the law of closed systems; it is sufficient to apply the laws of contraposition.

Theorem 10 and Axioms 10 and 11 form a closed system of three sentences. This is a consequence of the law of trichotomy; since between any two numbers we have exactly one of the relations $=$, $<$ and $>$, we see that the hypotheses of these three sentences, that is, the formulas:

$$y = z, \qquad y < z, \qquad y > z,$$

exhaust all possible cases, while their conclusions, that is, the formulas:

$$x + y = x + z, \qquad x + y < x + z, \qquad x + y > x + z,$$

exclude one another. (The law of trichotomy implies even more, which however is irrelevant for our purpose, namely, that the first three formulas do not only exhaust all possible cases but also exclude each other, and that the last three formulas do not only exclude each other but also exhaust all possible cases.) For the mere reason that the three statements form a closed system it is true that the converse theorems 11–13 must hold.

Numerous examples of closed systems can be found in elementary geometry; for instance, when examining the relative position of two circles, we have to deal with a closed system consisting of five sentences.

In conclusion it may be remarked that anyone who does not know the law of closed systems but tries to prove the converses of statements forming a system of this kind may mechanically apply the same mode of inference which we employed in the first proof of Theorem 11.

51. Consequences of the laws of monotony

Theorems 10 and 11 may be combined into one sentence:

$$y = z \quad \textit{if, and only if,} \quad x + y = x + z.$$

Similarly it is possible to combine Axioms 10 and 11 with Theorems 12 and 13. The theorems thus obtained may be denoted as the

LAWS OF EQUIVALENT TRANSFORMATION OF EQUATIONS AND INEQUALITIES by means of addition. The content of these theorems is sometimes described as follows: if the same number is added to both sides of an equation or inequality, without changing the equality or inequality sign, the resulting equation or inequality is equivalent to the original one (this formulation is, of course, not quite correct, since the sides of an equation or inequality are not numbers but expressions, to which it is not possible to add any numbers). The theorems mentioned here play an important role in the solution of equations and inequalities.

We will derive one more consequence from the theorems of monotony:

THEOREM 14. *If* $x + z < y + t$, *then* $x < y$ *or* $z < t$.

PROOF. Suppose the conclusion of the theorem is false; in other words, neither is x smaller than y nor is z smaller than t. From this it follows by the law of trichotomy that one of the two formulas:

$$x = y \quad \text{or} \quad x > y$$

and also one of the two formulas:

$$z = t \quad \text{or} \quad z > t$$

must hold. We thus have to discuss the following four possibilities:

(1) $x = y$ *and* $z = t$,

(2) $x = y$ *and* $z > t$,

(3) $x > y$ *and* $z = t$,

(4) $x > y$ *and* $z > t$.

Let us begin by considering the first case. If the two equations (1) are valid, we obtain, by Theorem 10:

$$z + x = z + y$$

from the first equation; and since, according to Axiom 7,

$$x + z = z + x \quad \text{and} \quad z + y = y + z$$

we may infer, by a twofold application of the law of transitivity for the relation of identity:

$$(5) \qquad x + z = y + z.$$

If now we apply Theorem 10 to the second of the equations (1), we obtain:

$$(6) \qquad y + z = y + t,$$

which, together with (5), yields:

$$(7) \qquad x + z = y + t.$$

By an entirely analogous inference—applying Axioms 4, 5, 10 and 11—any of the three remaining cases (2), (3) and (4) lead to the inequality:

$$(8) \qquad x + z > y + t.$$

One of the formulas (7) or (8) must therefore hold in any case. But since $x + z$ and $y + t$ are numbers (Axiom 6), it follows by the law of trichotomy that the formula:

$$x + z < y + t$$

cannot hold.

Thus, by assuming the conclusion to be false, we have arrived at an immediate contradiction to the hypothesis of the theorem. The assumption is therefore to be refuted, and we see that the conclusion does indeed follow from the hypothesis.

The argument just conducted is counted among the indirect proofs; apart from an inessential modification, it could be brought under the schema sketched in Section 49 in connection with the first proof of Theorem 11. Formally considered, however, the procedure of the argument is slightly different from the one followed in the proofs of Theorems 1 and 11. The inference has the following schema. In order to prove a sentence of the form of an implication, say, the sentence:

if p, then q,

we assume the conclusion of the sentence, that is, "*q*", to be false (and not the whole sentence); from this assumption, that is, from "*not q*", it is inferred that the hypothesis is false, that is, that "*not p*" holds. In other words, instead of demonstrating the sentence in question, a proof of the corresponding contrapositive sentence:

if not q, then not p

is given, and from this the validity of the original sentence is inferred. The basis for an inference of this kind is to be found in a law of sentential calculus to the effect that the truth of the contrapositive sentence always implies that of the original sentence (cf. Section 14).

Inferences of this form are very common in all mathematical disciplines; they constitute the most usual type of indirect proof.

52. Definition of subtraction; inverse operations

Our next task is to show how the notion of subtraction can be introduced into our considerations. With this aim in mind, we shall first prove the following theorem:

THEOREM 15. *For any two numbers y and z there is exactly one number x such that $y = z + x$.*

PROOF. Axiom 9 guarantees the existence of at least one number x satisfying the formula:

$$y = z + x.$$

We have to show that there is no more than one such number; in other words, that any two numbers u and v satisfying this formula are identical. Let, therefore,

$$y = z + u \quad \text{and} \quad y = z + v.$$

This implies at once (by the laws of symmetry and transitivity for the relation $=$):

$$z + u = z + v,$$

from which, by Theorem 11, we obtain:

$$u = v.$$

There is, thus, exactly one number x (cf. Section 20) for which

$$y = z + x, \qquad \text{q.e.d.}$$

This unique number x, of which the above theorem treats, is designated by the symbol:

$$y - z;$$

we read it, as usual, "*the difference of the numbers x and y*" or "*the result of subtracting the number z from the number y*". The precise definition of the notion of difference is as follows:

DEFINITION 2. *We say that* $x = y - z$ *if, and only if,* $y = z + x$.

An operation I is called a RIGHT INVERSE OF THE OPERATION O IN THE CLASS K if these two operations O and I fulfil the following condition:

for any elements x, y and z of the class K, we have: $x = y\,I\,z$ *if, and only if,* $y = z\,O\,x$.

The analogous concept of a LEFT INVERSE OF THE OPERATION O is defined similarly. If the operation O is commutative in the class K, its two inverses—the right and the left—coincide, and we can then simply speak of the INVERSE OF THE OPERATION O (or, also, of the INVERSE OPERATION OF O). In accordance with this terminology, Definition 2 expresses the fact that subtraction is the right inverse (or, simply, the inverse) of addition.

53. Definitions whose definiendum contains the identity sign

*Definition 2 exemplifies a kind of definition very common in mathematics. These definitions stipulate the meaning of a symbol designating either a single thing or an operation on a certain number of things (in other words, a function with a certain number of arguments). In every definition of this kind, the definiendum has the form of an equation:

$$x = \ldots;$$

on the right side of this equation, we have the symbol itself which was to be defined, or else a designatory function constructed out of the symbol to be defined and certain variables "y", "z", ..., according as the symbol in question designates a single thing or an operation on things. The definiens may be a sentential function of any form, which contains the same free variables as the definiendum, and which states that the thing x—together possibly with the things y, z, ...—satisfies such and such a condition.—Definition 2 establishes the meaning of a symbol which denotes an operation on two numbers. To give a different example of this type of definition, let us state the definition of the symbol "0" which designates a single number:

we say that $x = 0$ *if, and only if, for any number* y, *the formula:* $y + x = y$ *holds.*

A certain danger is connected with definitions of the type under consideration; for if one does not proceed with sufficient caution in laying down such definitions, one can easily find oneself confronted with a contradiction. A concrete example will make this clear.

Let us leave, for the moment, our present investigations, and assume that in arithmetic we have already the symbol of multiplication at our disposal and that, with its help, we want to define the symbol of division. For this purpose we proceed to lay down the following definition, which is modelled precisely after Definition 2:

we say that $x = y : z$ *if, and only if,* $y = z \cdot x$.

If now, in this definition, we replace both "y" and "z" by "0", and "x" first by "1" and then by "2", and if we observe that we have the formulas:

$$0 = 0 \cdot 1 \quad \text{and} \quad 0 = 0 \cdot 2,$$

we obtain at once:

$$1 = 0:0 \quad \text{and} \quad 2 = 0:0.$$

But since two things equal to the same thing are equal to each other, we arrive at:

$$1 = 2,$$

which is obviously nonsense.

It is not hard to exhibit the reason for this phenomenon. Both in Definition 2 and in the definition of the quotient considered here, the definiens has the form of a sentential function with three free variables "x", "y" and "z". To each such sentential function there corresponds a three-termed relation holding between the numbers x, y and z if, and only if, these numbers satisfy that sentential function (cf. Section 27); and it is just the aim of the definition to introduce a symbol designating this relation. But if one gives the definiendum the form:

$$x = y - z \quad \text{or} \quad x = y : z,$$

one assumes in advance that this relation is functional (and hence an operation, or a function, cf. Section 34), and that therefore, to any two numbers y and z, there is at most one number x standing to them in the relation in question. The fact that the relation is functional, however, is not at all evident from the beginning, and it must first be established. This we did in the case of Definition 2; but we failed to do so in the case of the definition of the quotient, and we would indeed have been unable to do so, simply because the relation in question ceases to be functional in a certain exceptional case: for, if

$$y = 0 \quad \text{and} \quad z = 0,$$

there exist infinitely many numbers x for which

$$y = z \cdot x.$$

If, therefore, one wants to formulate the definition of the quotient in the above form without introducing contradictions, one has to take care that the case is excluded where both numbers y and z are 0,—for instance, by inserting an additional condition in the definiens.

The above considerations lead us to the following conclusion. Every definition of the type of Definition 2 should be preceded by a theorem corresponding exactly to Theorem 15, that is to say, a theorem to the effect that there is but one number x which satisfies the definiens. (The question arises whether it is relevant if there is exactly one number x, or whether it is sufficient that there is at most one such number. A discussion of this rather difficult problem will be omitted here.)*

54. Theorems on subtraction

On the basis of Definition 2 and the laws of addition we can without difficulty prove the fundamental theorems of the theory of subtraction, such as the law of performability, the laws of monotony, and the laws of equivalent transformation of equations and inequalities by means of subtraction. Those theorems also belong here which make possible the transformation of so-called algebraic sums, that is, of expressions consisting of numerical constants and variables, separated by "+" and "−" signs as well as parentheses (the latter often being omitted in accordance with special rules to this effect). The following theorem may serve as an example of the last-named category:

THEOREM 16. $x + (y - z) = (x + y) - z.$

PROOF. To y and z, according to Axiom 9, there corresponds a number u such that

(1) $$y = z + u;$$

this implies, by Definition 2,

(2) $$u = y - z.$$

From the commutative law we have:

$$x + y = y + x.$$

On account of (1), "y" may here be replaced by "$z + u$" on the right side, so that we obtain:

(3) $$x + y = (z + u) + x.$$

From Theorem 9, on the other hand, it follows that:

(4) $$z + (x + u) = (z + u) + x.$$

But since two numbers equal to the same number are equal to each other, we can infer from (3) and (4):

(5) $$x + y = z + (x + u).$$

Now, since $x + u$ and $x + y$ are numbers (by Axiom 6), we may substitute "$x + u$" and "$x + y$" for "x" and "y" in Definition 2.

(5) shows that the definiens is then satisfied, and hence the definiendum must also hold:

$$x + u = (x + y) - z.$$

If now, in view of (2), we replace "u" by "$y - z$" in this last equation, we finally arrive at:

$$x + (y - z) = (x + y) - z, \qquad \text{q.e.d.}$$

Having gotten this far, we now terminate the construction of our fragment of arithmetic.

Exercises

1. Consider the following three systems, each consisting of a certain set, two relations and one operation:

(a) the set of all numbers, the relations $\leqq$ and $\geqq$, the operation of addition;

(b) the set of all numbers, the relations $<$ and $>$, the operation of multiplication;

(c) the set of all positive numbers, the relations $<$ and $>$, the operation of multiplication.

Determine which of these systems are models of the system of Axioms 1–11 (cf. Section 37).

2. Consider an arbitrary straight line, to which we will refer as the number line; let the points on this line be denoted by the letters "X", "Y", "Z", ... On the number line we choose a fixed initial point O and a unit point U distinct from O. Now let X and Y be any two distinct points on our line. We consider the two half-lines, one beginning at O and going through U, the other beginning at X and going through Y. We shall say that the point X precedes the point Y, in symbols:

$$X \circledless Y,$$

if, and only if, either the two half-lines are identical or one of them —no matter which—is a part of the other. In the same situation we shall also say that the point Y succeeds the point X, written:

$$Y \circledgtr X.$$

The point Z is called the sum of the points X and Y if it fulfils the following conditions: (i) the segment OX is congruent to the segment YZ; (ii) if $O \circledcirc X$, then $Y \circledcirc Z$, but if $O \circledcirc X$, then $Y \circledcirc Z$. The sum of the points X and Y is denoted by:

$$X \oplus Y.$$

Show by means of the theorems of geometry that the set of all points of the number line (that is, more simply, the number line itself), the relations ⩹ and ⩺, and the operation $\oplus$, together form a model of the axiom system adopted by us, and that, therefore, this system has an interpretation within geometry.

3. Let us consider four operations **A, B, G** and **L** which—like addition—correlate a third number with any two numbers. As the result of the operation **A** on the numbers x and y we always consider the number x, and as the result of the operation **B** the number y:

$$x \mathbf{A} y = x, \qquad x \mathbf{B} y = y.$$

By the symbols "$x \mathbf{G} y$" and "$x \mathbf{L} y$" we denote that of the two numbers x and y which is not less than or not greater than the other, respectively; we thus have:

$$x \mathbf{G} y = x \quad \text{and} \quad x \mathbf{L} y = y \quad \text{in case that} \quad x \geqq y;$$

$$x \mathbf{G} y = y \quad \text{and} \quad x \mathbf{L} y = x \quad \text{in case that} \quad x \leqq y.$$

Which of the properties discussed in Section 47 belong to these four operations? Is the set of all numbers a group and, in particular, an Abelian group with respect to any of these operations?

4. Let **C** be the class of all point sets, that is, of all geometrical configurations. Are the addition and multiplication of sets (as defined in Section 25) performable, commutative, associative and invertible in the class **C**? Is, therefore, the class **C** a group and, in particular, an Abelian group with respect to any of these operations?

5. Show that the set of all numbers is not an Abelian group with respect to multiplication, but that every one of the following sets is an Abelian group with respect to that operation:

(a) the set of all numbers different from 0;

(b) the set of all positive numbers;

(c) the set consisting of the two numbers 1 and -1.

6. Consider the set **S** consisting of the two numbers 0 and 1, and let the operation $\oplus$ on the elements of this set be defined by the following formulas:

$$0 \oplus 0 = 1 \oplus 1 = 0,$$

$$0 \oplus 1 = 1 \oplus 0 = 1.$$

Determine whether the set **S** is an Abelian group with respect to the operation $\oplus$.

7. Consider the set **S** consisting of the three numbers 0, 1 and 2. Define an operation $\oplus$ on the elements of this set, so that the set **S** will be an Abelian group with respect to this operation.

8. Prove that no set consisting of two or three different numbers can be an Abelian group with respect to addition. Is there a set consisting of one single number that forms an Abelian group with respect to addition?

9. Derive the following theorems from Axioms 6–8:

(a) $x + (y + z) = (z + x) + y;$

(b) $x + [y + (z + t)] = (t + y) + (x + z).$

10. How many expressions can be obtained from each of these expressions:

$$x + (y + z), \quad x + [y + (z + t)], \quad x + \{y + [z + (t + u)]\}$$

if they are transformed solely on the basis of Axioms 6–8?

11. Formulate the general definition of left monotony of an operation O with respect to a relation R.

12. On the basis of the axioms adopted by us and the theorems derived from them, prove that addition is a monotonic operation with respect to the relations $\neq$, $\leqq$ and $\geqq$.

13. Is multiplication a monotonic operation with respect to the relations $<$ and $>$

(a) in the set of all numbers,

(b) in the set of all positive numbers,

(c) in the set of all negative numbers?

14. Which of the operations defined in Exercise 3 are monotonic with respect to the relations $=$, $<$, $>$, $\neq$, $\leqq$ and $\geqq$?

15. Are the addition and multiplication of classes monotonic with respect to the relation of inclusion? Or with respect to any of the other relations among classes discussed in Section 24?

16. Derive from our axioms the following theorem:

$$\textit{if} \quad x < y \quad \textit{and} \quad z < t, \quad \textit{then} \quad x + z < y + t.$$

Replace in this sentence the symbol "$<$" in turn by "$>$", "$=$", "$\neq$", "$\leqq$" and "$\geqq$", and examine whether the sentences obtained in this way are true.

17. Give examples of closed systems of sentences within arithmetic and geometry.

18. Derive the following theorems from our axioms:

(a) *if* $x + x = y + y$, *then* $x = y$;

(b) *if* $x + x < y + y$, *then* $x < y$;

(c) *if* $x + x > y + y$, *then* $x > y$.

Hint: Prove the converse sentences first (using the results of Exercise 16), and show that they form a closed system.

*19. If a theorem is derivable from Axioms 6–9 alone, it can be extended to arbitrary Abelian groups, since every class K which forms an Abelian group with respect to an operation O constitutes, together with this operation, a model of Axioms 6–9 (cf. Sections 37 and 38). This applies, in particular, to Theorem 11 (in view of the second proof of this theorem), and we have the following general group-theoretical theorem:

every class K which is an Abelian group with respect to the operation O satisfies the following condition:

$$\text{if } x \in K, \quad y \in K, \quad z \in K \quad \text{and} \quad x\,O\,y = x\,O\,z, \quad \text{then} \quad y = z.$$

Give a strict proof of this theorem.

Show, on the other hand, that Theorem (a) of Exercise 18 cannot be extended to arbitrary Abelian groups, by exhibiting an example of a class K and an operation O with these properties: (i) the class K is an Abelian group with respect to the operation O, and (ii) there exist two distinct elements x and y of the class K for which $x\,O\,x = y\,O\,y$ (cf. Exercise 6). Consequently, is it possible to derive Theorem (a) from Axioms 6–9 alone?

20. Transform the proof of Theorem 14 in such a manner that it conforms to the schema sketched in Section 49 in connection with the first proof of Theorem 11.

21. May the operation of division be said to be the inverse of multiplication in the set of all numbers?

22. Do the operations mentioned in Exercises 3 and 4 possess inverses (in the set of all numbers, or in the class of all geometrical configurations)?

23. What operations are the left and the right inverses of subtraction (in the set of all numbers)?

*24. In Section 53, the definition of the symbol "0" was stated by way of an example. In order to be certain that this definition does not lead to a contradiction, it should be preceded by the following theorem:

there exists exactly one number x such that, for any number y, we have: $y + x = y$.

Prove this theorem on the basis of Axioms 6–9 alone.

25. Formulate the sentences which assert that subtraction is performable, commutative, associative, right- and left-invertible, and right- and left-monotonic with respect to the relation *less than*. Which of these sentences are true? Prove those for which this is the case, using our axioms and Definition 2 of Section 52.

26. Derive the following theorems from our axioms and Definition 2:

(a) $x - (y + z) = (x - y) - z,$

(b) $x - (y - z) = (x - y) + z,$

(c) $x + y = x - [(x - y) - x].$

***27.** Using the law of performability for subtraction and Theorem (c) of the preceding exercise, prove the following theorem:

for a set K of numbers to be an Abelian group with respect to addition, it is necessary and sufficient that the difference of any two numbers of the set K also belongs to the set K (i.e. *that the formulas* $x \in K$ *and* $y \in K$ *always imply* $x - y \in K$).

Use this theorem in order to find examples of sets of numbers that are Abelian groups with respect to addition.

28. Write in logical symbolism all axioms, definitions and theorems given in the last two chapters.

Hint: Before formulating Theorem 15 in symbols, put it in an equivalent form in which the numerical quantifiers have been eliminated by virtue of the explanations given in Section 20.

· IX ·

METHODOLOGICAL CONSIDERATIONS ON THE CONSTRUCTED THEORY

55. Elimination of superfluous axioms in the original axiom system

The two preceding chapters were devoted to an outline of the foundations of an elementary mathematical theory which constitutes a fragment of arithmetic. In the present chapter we shall proceed to considerations of a methodological nature, concerning the system of axioms and primitive terms upon which that theory is based.

We shall begin with concrete examples illustrating the remarks of Section 39 concerning such problems as arbitrariness in the selection of axioms and primitive terms, the possible omission of superfluous axioms, and so on.

Let us start out with the question whether our system of Axioms 1–11—it will briefly be referred to as SYSTEM $\mathfrak{A}$—possibly contains any superfluous axioms, that is, axioms which can be derived from the remaining axioms of the system. We shall see at once that it is easy to answer this question, and, moreover, affirmatively. In fact, we have:

Three of the axioms of System $\mathfrak{A}$, *namely, one of the Axioms* 4 *or* 5, *Axiom* 6, *and one of the Axioms* 10 *or* 11, *can be derived from the remaining axioms.*

PROOF. We show first that

(I) *either of the Axioms* 4 *or* 5 *can be derived from the other with the help of Axioms* 1–3.

In fact, we observe that the proof of Theorem 3 was based exclusively—whether directly or indirectly—upon Axioms 1–3. If, on the other hand, we already have Theorem 3 at our disposal, we may derive Axiom 5 from Axiom 4 (or vice versa) by the following mode of inference:

If

$$x > y \quad \textit{and} \quad y > z,$$

then, by Theorem 3,

$$y < x \quad \textit{and} \quad z < y;$$

hence, by applying Axiom 4 (with "x" having been replaced by "z", and "z" by "x"), we obtain:

$$z < x,$$

which, again by Theorem 3, implies:

$$x > z,$$

and this is the conclusion of Axiom 5.

Similarly it can be shown that:

(II) *either of the Axioms* 10 *or* 11 *can be derived from the other with the help of Axioms* 1–3.

Finally, we have:

(III) *Axiom* 6 *can be derived from Axioms* 7–9.

*The proof of this latter assertion is not quite simple and resembles the second proof of Theorem 11. Two arbitrary numbers x and y are given; by a fourfold application of Axiom 9, four new numbers u, w, z, and v are introduced one by one, satisfying the following formulas:

(1) $$y = y + u,$$

(2) $$u = x + w,$$

(3) $$y = w + z,$$

(4) $$z = y + v.$$

From (1) we have, by the commutative law,

$$y = u + y;$$

combining this equation with (4) and arguing as in the case of the proof of Theorem 11, we obtain, by the associative law,

$$z = u + z. \tag{5}$$

From (5) and (2) we obtain:

$$z = (x + w) + z,$$

and hence, again by the associative law,

$$z = x + (w + z)$$

which, in view of (3), yields:

$$z = x + y. \tag{6}$$

We have thus shown that, for any two numbers x and y, there exists a number z for which (6) holds; and this is just the content of Axiom 6.

It might be added that the mode of inference sketched above applies, not only to addition, but—in accordance with the general remarks of Sections 37 and 38—also to any other operation; an operation O which is commutative, associative and right-invertible in a class K is also performable in that class, and the class K, therefore, forms an Abelian group with respect to the operation O (cf. Section 47).*

We have seen now that System $\mathfrak{A}$ contains at least three axioms which are superfluous and may therefore be omitted. Consequently, System $\mathfrak{A}$ may be replaced by the system consisting of the following eight axioms:

Axiom 1$^{(1)}$. *For any numbers x and y we have: $x = y$ or $x < y$ or $x > y$.*

Axiom 2$^{(1)}$. *If $x < y$, then $y \nless x$.*

Axiom 3$^{(1)}$. *If $x > y$, then $y \ngtr x$.*

Axiom 4$^{(1)}$. *If $x < y$ and $y < z$, then $x < z$.*

AXIOM $5^{(1)}$. $x + y = y + x$.

AXIOM $6^{(1)}$. $x + (y + z) = (x + y) + z$.

AXIOM $7^{(1)}$. *For any numbers x and y there exists a number z such that* $x = y + z$.

AXIOM $8^{(1)}$. *If* $y < z$, *then* $x + y < x + z$.

We shall refer to this axiom system as SYSTEM $\mathfrak{A}^{(1)}$, and we now have the following result:

Systems $\mathfrak{A}$ and $\mathfrak{A}^{(1)}$ are equipollent.

In comparison with the original system, the new simplified axiom system has certain shortcomings, both from the esthetic and the didactical points of view. It is no longer symmetric with respect to the two primitive symbols "$<$" and "$>$", certain properties of the relation $<$ being accepted without proof, while quite analogous properties of the relation $>$ have first to be demonstrated. Also, in the new system, Axiom 6 is missing which was of a very elementary and intuitively evident character, while its derivation from the axioms contained in System $\mathfrak{A}^{(1)}$ offers some difficulties.

56. Independence of the axioms of the simplified system

The question now arises whether there are any further superfluous axioms contained in System $\mathfrak{A}^{(1)}$. It will turn out that this is not the case:

$\mathfrak{A}^{(1)}$ is a system of mutually independent axioms.

In order to establish the methodological statement just formulated, we employ the method of proof by interpretation, which has already been used in a particular case in Section 37.

We are to show that no axiom of System $\mathfrak{A}^{(1)}$ is derivable from the remaining axioms of this system. Let us consider, for example, Axiom $2^{(1)}$. Suppose we replace the symbol "$<$" in the axioms of System $\mathfrak{A}^{(1)}$ throughout by "$\leqq$", without altering the axioms in any other way. As a result of this transformation, no axiom, with the exception of Axiom $2^{(1)}$, loses its validity; in fact, Axioms $3^{(1)}$, $5^{(1)}$, $6^{(1)}$ and $7^{(1)}$, since they do not contain the symbol "$<$", are left unaltered, and Axioms $1^{(1)}$, $4^{(1)}$ and $8^{(1)}$ go over into certain arithmetical theorems whose proofs on the basis of System $\mathfrak{A}$ or $\mathfrak{A}^{(1)}$ and the Definition 1 of the symbol "$\leqq$" (cf. Section 46) present no difficulties. It may, therefore, be asserted that the set **N** of all numbers, the relations $\leqq$ and $>$, and the operation of addition form a model of the Axioms $1^{(1)}$ and $3^{(1)}$–$8^{(1)}$; the system of these seven axioms has thus found a new interpretation within arithmetic. On the other hand, it can be seen at once that the sentence resulting from Axiom $2^{(1)}$ by the transformation is false, for its negation can easily be proved in arithmetic; the formula:

$$x \leqq y$$

does not always exclude:

$$y \leqq x,$$

for there are numbers x and y simultaneously satisfying the two inequalities:

$$x \leqq y \quad \text{and} \quad y \leqq x$$

(this, of course, is the case if, and only if, x and y are equal). If, therefore, one believes in the consistency of arithmetic (cf. Section 41), one has to accept the fact that the sentence obtained from Axiom $2^{(1)}$ is not a theorem of this discipline. And from this it follows that Axiom $2^{(1)}$ is not derivable from the remaining axioms of System $\mathfrak{A}^{(1)}$; for otherwise this axiom could not fail to be valid in the case of any interpretation in which the other axioms hold (cf. analogous considerations in Section 37).

By using the same method of reasoning but by applying other, suitable interpretations, we can obtain the same result for any of the other axioms.

*In general, the method of proof by interpretation can be described as follows. It is a question of showing that some sentence A is not a consequence of a certain system $\mathfrak{S}$ of axioms or other statements of a given deductive theory. For this purpose, we consider an arbitrary deductive theory $\mathfrak{T}$ of which we assume that it is consistent (it may, in particular, be the same theory to which the statements of the system $\mathfrak{S}$ belong). We then try to find, within this theory, an interpretation of the system $\mathfrak{S}$ of such a kind that not the sentence A itself but its negation becomes a theorem (or possibly an axiom) of the theory $\mathfrak{T}$. If we are successful in doing so, we may apply the law of deduction stated in Section 38. As we know, it follows from this law that, if the sentence A could be derived from the statements of the system $\mathfrak{S}$, it would remain valid for any interpretation of this system. Consequently, the very fact of the existence of an interpretation of $\mathfrak{S}$ for which A is not valid is a proof that this sentence cannot be derived from the system $\mathfrak{S}$. More strictly speaking, it is a proof of the conditional sentence:

if the theory $\mathfrak{T}$ is consistent, then the sentence A *cannot be derived from the statements of the system* $\mathfrak{S}$.

The reason why we must include the hypothesis that the theory $\mathfrak{T}$ is consistent is easily seen. For otherwise the theory $\mathfrak{T}$ could contain two contradictory sentences among its axioms and theorems, and we could not conclude that $\mathfrak{T}$ did not contain the sentence A (or rather the interpretation of A), from the mere fact that $\mathfrak{T}$ did contain the negation of A; thus our argument would no longer be valid.

In order to arrive, in the above way, at an exhaustive proof of the independence of a given axiom system, the method described has to be applied as many times as there are axioms in the system in question; each axiom in turn is taken as the sentence A, while $\mathfrak{S}$ consists of the remaining axioms of the system.*

57. Elimination of superfluous primitive terms and subsequent simplification of the axiom system; concept of an ordered Abelian group

We return once more to the axiom system $\mathfrak{A}^{(1)}$. Since this system is independent, it does not permit of any further simplification by the omission of superfluous axioms. Nevertheless, a simplification can be achieved in a different way. For it turns out that the primitive terms of System $\mathfrak{A}^{(1)}$ are not mutually independent. In fact, either one of the two symbols "$<$" and "$>$" may be stricken from the list of primitive terms, and then it can be defined in terms of the other. This is easily seen from Theorem 3; on account of its form, this theorem may be considered as a definition of the symbol "$>$" by means of the symbol "$<$", and if in this theorem we exchange the two sides of the equivalence, we may look upon it as a definition of the symbol "$<$" by means of the symbol "$>$". (In either case it is desirable to have the phrase "*We say that*" precede the theorem; cf. Section 11.) From the didactical point of view, this reduction of the primitive terms might provoke certain objections; for the terms "$<$" and "$>$" are equally clear in their meaning and the relations denoted by them possess entirely analogous properties, so that it would appear slightly artificial to consider one of these terms immediately comprehensible while the other has first to be defined with the help of the first. But these objections carry little conviction.

If now, in disregard of any didactical considerations, we resolve to eliminate one of the symbols in question from the list of primitive terms, the task arises of giving our axiom system a form in which no defined terms occur in it. (This is a methodological postulate, by the way, which in practice is frequently disregarded; in geometry, especially, the axioms are usually formulated with the help of defined terms in order to enhance their simplicity and evidence). This task does not present any difficulties; we simply replace in the axiom system $\mathfrak{A}^{(1)}$ every formula of the type:

$$x > y$$

by the formula:

$$y < x$$

which, by Theorem 3, is equivalent to it. It is then easily seen that Axiom 1 may be replaced by the law of connexity, i.e. Theorem 4, since each follows from the other on the basis of general laws of logic (of sentential calculus, that is); Axiom 3 now becomes a simple substitution of Axiom 2, and may hence be omitted altogether. In this way we arrive at the system consisting of the following seven axioms:

AXIOM $1^{(2)}$. *If* $x \neq y$, *then* $x < y$ *or* $y < x$.

AXIOM $2^{(2)}$. *If* $x < y$, *then* $y \nless x$.

AXIOM $3^{(2)}$. *If* $x < y$ *and* $y < z$, *then* $x < z$.

AXIOM $4^{(2)}$. $x + y = y + x$.

AXIOM $5^{(2)}$. $x + (y + z) = (x + y) + z$.

AXIOM $6^{(2)}$. *For any numbers x and y there exists a number z such that* $x = y + z$.

AXIOM $7^{(2)}$. *If* $y < z$, *then* $x + y < x + z$.

This axiom system, called SYSTEM $\mathfrak{A}^{(2)}$, is thus equipollent to either of the two former Systems $\mathfrak{A}$ and $\mathfrak{A}^{(1)}$. However, in saying this, we commit one inexactitude; for it is impossible to derive from the axioms of System $\mathfrak{A}^{(2)}$ those sentences of Systems $\mathfrak{A}$ or $\mathfrak{A}^{(1)}$ which contain the symbol "$>$", unless System $\mathfrak{A}^{(2)}$ is enlarged by adding the definition of this symbol. We may, as we know, give this definition the following form:

DEFINITION $1^{(2)}$. *We say that* $x > y$ *if, and only if,* $y < x$.

We also know that this last sentence can be proved on the basis of Systems $\mathfrak{A}$ or $\mathfrak{A}^{(1)}$, if it is treated, not as a definition, but as an ordinary theorem (omitting, in this case, the initial phrase "*We say that*"). The fact of the equipollence of the three systems in questions can now be formulated as follows:

System $\mathfrak{A}^{(2)}$ together with Definition $1^{(2)}$ is equipollent to each of the Systems $\mathfrak{A}$ and $\mathfrak{A}^{(1)}$.

An equally cautious mode of formulation is indicated whenever two axiom systems are compared which, though equipollent, contain, at least partly, different primitive terms.

The axiom system $\mathfrak{A}^{(2)}$ is distinguished advantageously by the simplicity of its structure. The first three axioms concern the relation *less than*, and together they assert that the set **N** is ordered by this relation; the next three axioms are concerned with addition, and they assert that the set **N** is an Abelian group with respect to addition; the last axiom finally—the law of monotony—states a certain dependence between the relation *less than* and the operation of addition. A class K is said to be an ORDERED ABELIAN GROUP WITH RESPECT TO THE RELATION R AND THE OPERATION O if (i) the class K is ordered by the relation R, (ii) the class K is an Abelian group with respect to the operation O, and (iii) the operation O is monotonic in the class K with respect to the relation R. In accordance with this terminology we can say that the set of numbers is characterized by the axiom system $\mathfrak{A}^{(2)}$ as an ordered Abelian group with respect to the relation *less than* and the operation of addition.

The following facts concerning System $\mathfrak{A}^{(2)}$ can be established:

System $\mathfrak{A}^{(2)}$ is an independent axiom system, and moreover, all its primitive terms, namely "**N**", "$<$" *and* "$+$", *are mutually independent.*

We omit the proof of this statement. We remark only that, in order to establish the mutual independence of the primitive terms, one has again to apply the method of proof by interpretation, which in this case, however, assumes a more involved form; lack of space prevents us from going into the modifications of that method which would be required for this purpose.

58. Further simplification of the axiom system; possible transformations of the system of primitive terms

System $\mathfrak{A}^{(2)}$ can obviously be replaced by any system of sentences equipollent to it. We will here give a particularly simple example of such a system, which may be called SYSTEM $\mathfrak{A}^{(3)}$, and which contains the same primitive terms as $\mathfrak{A}^{(2)}$. It consists of only five sentences:

AXIOM $1^{(3)}$. *If* $x \neq y$, *then* $x < y$ *or* $y < x$.

AXIOM $2^{(3)}$. *If* $x < y$, *then* $y \nless x$.

AXIOM $3^{(3)}$. $x + (y + z) = (x + z) + y$.

AXIOM $4^{(3)}$. *For any numbers x and y there exists a number z such that* $x = y + z$.

AXIOM $5^{(3)}$. *If* $x + z < y + t$, *then* $x < y$ *or* $z < t$.

We shall show that

Systems $\mathfrak{A}^{(2)}$ and $\mathfrak{A}^{(3)}$ are equipollent.

PROOF. We observe, first of all, that all the axioms of System $\mathfrak{A}^{(3)}$ are either contained in System $\mathfrak{A}$ (thus, Axiom $2^{(3)}$ coincides with Axiom 2, and Axiom $4^{(3)}$ with Axiom 9), or else have been proved on its basis (Axioms $1^{(3)}$, $3^{(3)}$ and $5^{(3)}$ are coincident with Theorems 4, 9 and 14, respectively). But since the axiom systems $\mathfrak{A}$ and $\mathfrak{A}^{(2)}$ are equipollent, as we know from Section 57 (Definition $1^{(2)}$, after all, can always be added to System $\mathfrak{A}^{(2)}$), we may conclude that all the sentences of System $\mathfrak{A}^{(3)}$ can be proved on the basis of System $\mathfrak{A}^{(2)}$. It remains to derive those sentences of System $\mathfrak{A}^{(2)}$ from the axioms of System $\mathfrak{A}^{(3)}$ which are absent in $\mathfrak{A}^{(3)}$, that is Axioms $3^{(2)}$, $4^{(2)}$, $5^{(2)}$ and $7^{(2)}$. This task is not quite so simple.

*We begin with Axioms $4^{(2)}$ and $5^{(2)}$.

(I) *Axiom $4^{(2)}$ can be derived from the axioms of System $\mathfrak{A}^{(3)}$.*

For, given two numbers x and y, we can apply Axiom $4^{(3)}$ (with "x" put in place of "y", and vice versa); there is, therefore, a number z such that

(1) $$y = x + z.$$

If now, in Axiom $3^{(3)}$, we replace "y" by "x", we obtain:

(2) $$x + (x + z) = (x + z) + x.$$

In view of (1), "$x + z$" may here be replaced by "y" on both sides, and we arrive at Axiom $4^{(2)}$:

$$x + y = y + x$$

(II) *Axiom $5^{(2)}$ can be derived from the axioms of System $\mathfrak{A}^{(3)}$.*

In fact, by Axiom $3^{(3)}$ we have (if "y" is substituted for "z", and conversely):

$$x + (z + y) = (x + y) + z;$$

on account of the commutative law, which has already been derived by (I), we may here replace "$z + y$" by "$y + z$", and we obtain Axiom $5^{(2)}$:

$$x + (y + z) = (x + y) + z.$$

In order to facilitate the derivation of Axioms $3^{(2)}$ and $7^{(2)}$, we shall first show how some of the axioms and theorems stated in the preceding chapters may be proved on the basis of System $\mathfrak{A}^{(3)}$.

(III) *Theorem* 1 *can be derived from the axioms of System* $\mathfrak{A}^{(3)}$.

We merely observe that the proof of Theorem 1 given in Section 44 is based exclusively upon Axiom 2, which in turn coincides with Axiom $2^{(3)}$ of System $\mathfrak{A}^{(3)}$.

(IV) *Axiom* 6 *can be derived from the axioms of System* $\mathfrak{A}^{(3)}$.

In fact, we saw in Section 55 that Axiom 6 can be deduced from Axioms 7, 8 and 9. Axioms 7 and 8 are the same as Axioms $4^{(2)}$ and $5^{(2)}$, and can therefore, by (I) and (II), be derived from the axioms of System $\mathfrak{A}^{(3)}$. Axiom 9, on the other hand, occurs as Axiom $4^{(3)}$ in System $\mathfrak{A}^{(3)}$. Hence, Axiom 6 is a consequence of the axioms of $\mathfrak{A}^{(3)}$.

(V) *Theorem* 11 *can be derived from the axioms of System* $\mathfrak{A}^{(3)}$.

In the second proof of Theorem 11, as given in Section 49, only Axioms 7, 8 and 9 were used. Theorem 11 is therefore derivable from the axioms of System $\mathfrak{A}^{(3)}$ for the same reason as Axiom 6; see (IV).

(VI) *Theorem* 12 *can be derived from the axioms of System* $\mathfrak{A}^{(3)}$.

For, suppose the hypothesis of Theorem 12 holds:

$$x + y < x + z;$$

we apply Axiom $5^{(3)}$, having replaced "z", "y" and "t" by "y", "x" and "z", respectively. It follows that one of the formulas:

$$x < x \quad \text{or} \quad y < z$$

must hold; the first possibility has to be rejected because it contradicts Theorem 1 which has already been shown to be derivable from System $\mathfrak{A}^{(3)}$,—cf. (III). Hence the conclusion of Theorem 12 must hold:

$$y < z.$$

(VII) *Axiom* $3^{(2)}$ *can be derived from the axioms of System* $\mathfrak{A}^{(3)}$.

Let us assume the hypothesis of Axiom $3^{(2)}$, that is, the formulas:

$$x < y \tag{1}$$

and

(2) $$y < z.$$

If now we had:

$$y + x = y + z,$$

it would follow by Theorem 11, which has already been derived by (V), that

$$x = z.$$

Thus in (1) "x" might be replaced by "z", which would lead to:

$$z < y.$$

This inequality would contradict (2) by virtue of Axiom $2^{(3)}$ and must therefore be rejected. We thus have:

(3) $$y + x \neq y + z.$$

Since, by Axiom 6, $y + x$ and $y + z$ are numbers, we may, by Axiom $1^{(3)}$, infer from (3) that one of these two cases must hold:

(4) $$y + x < y + z \quad or \quad y + z < y + x.$$

Considering the second of the formulas (4), we may replace in it "$y + x$" by "$x + y$" by virtue of Axiom $4^{(2)}$ which has already been derived; we thus arrive at:

$$y + z < x + y.$$

To this formula we apply Axiom $5^{(3)}$, where we replace "x" and "t" by "y", and "y" by "x". We arrive in this way at the following consequence:

$$y < x \quad or \quad z < y.$$

But this has to be rejected, since, on account of Axiom $2^{(3)}$, it contradicts (1) and (2) which constitute the hypothesis of Axiom $3^{(2)}$. We therefore return to the first of the formulas (4) and apply Theorem 12 derived above by (VI), with "x" replaced by "y", and conversely; we obtain thus:

$$x < z,$$

and this is the conclusion of Axiom $3^{(2)}$.

(VIII) *Axiom $7^{(2)}$ can be derived from the axioms of System $\mathfrak{A}^{(3)}$.*

The procedure is here similar to the one just applied, but much simpler. We assume the hypothesis of Axiom $7^{(2)}$:

(1) $$y < z.$$

If now we had:

$$x + y = x + z,$$

it would follow by Theorem 11 that

$$y = z;$$

in (1) we might therefore replace "y" by "z" and arrive at a contradiction to Theorem 1, derived above by (III). Hence we must have:

$$x + y \neq x + z,$$

from which, by Axiom $1^{(3)}$, it follows that

(2) $$x + y < x + z \quad or \quad x + z < x + y.$$

In view of Theorem 12, the second of these inequalities leads to:

$$z < y,$$

but this contradicts our hypothesis (1) by virtue of Axiom $2^{(3)}$. Consequently we have to accept the first of the inequalities (2):

$$x + y < x + z,$$

and this is the conclusion of Axiom $7^{(2)}$.*

We have seen in this manner that all sentences of System $\mathfrak{A}^{(2)}$ are consequences of System $\mathfrak{A}^{(3)}$, and conversely; the two axiom systems $\mathfrak{A}^{(2)}$ and $\mathfrak{A}^{(3)}$ are thus really established as equipollent.

System $\mathfrak{A}^{(3)}$, no doubt, is simpler than System $\mathfrak{A}^{(2)}$, and hence still simpler than Systems $\mathfrak{A}$ or $\mathfrak{A}^{(1)}$. Particularly interesting is a comparison between Systems $\mathfrak{A}$ and $\mathfrak{A}^{(3)}$; as a result of the successive reductions that have been carried out, the original number of axioms has been diminished by more than one half. On the other hand, it should be noted that some of the sentences of System $\mathfrak{A}^{(3)}$ (namely, Axioms $3^{(3)}$ and $5^{(3)}$) are less natural and simple than the axioms of the other systems, and also that the proofs of some, even very elementary, theorems are here comparatively more difficult and involved than on the basis of those other systems.

Just like a system of axioms, a system of primitive terms may be replaced by any equipollent system. This applies, in particular, to the system of the three terms "**N**", "$<$" and "$+$" which occur as the only primitive terms in the axioms last considered. If, for instance, in this system we replace the symbol "$<$" by "$\leqq$", we obtain an equipollent system; for the second of these symbols was defined in terms of the first, and Theorem 8 tells us how the first may be defined by means of the second. But such a transformation of the system of primitive terms would be in no way advantageous; in particular, it would contribute nothing to a simplification of the axioms, and to the reader, who is possibly more familiar with the symbol "$<$" than with the symbol "$\leqq$", it might even appear rather artificial. Another equipollent system can be obtained by replacing in the original system the symbol "$+$" by "$-$"; but, again, this transformation would not be at all expedient. In conclusion we should note that other systems of primitive terms are known which are equipollent to the system in question and consist of but two terms.

59. Problem of the consistency of the constructed theory

We shall now briefly touch on some other methodological problems concerning the fragment of arithmetic considered above; these are the problems of consistency and of completeness (cf. Section 41). Since it is quite irrelevant whether we refer our remarks to one or another of several equipollent axiom systems, we shall now always speak of System $\mathfrak{A}$.

If we believe in the consistency of the whole of arithmetic (and this assumption has been made previously and will be made again in our further considerations), then we must all the more accept the fact that

The mathematical theory based on System $\mathfrak{A}$ is consistent.

But while the attempts to give a strict proof of the consistency of the whole of arithmetic have met with essential difficulties (cf. Section 41), a proof of this kind for System $\mathfrak{A}$ is not only possible but even comparatively simple. One reason for this is the fact that the variety of theorems which can be derived from the axiom system 𝔄 is very small indeed; it is, for instance, not possible to give, on its basis, an answer to the very elementary question as to whether any numbers exist at all. This circumstance facilitates considerably the proof of the fact that the part of arithmetic considered does not contain a single pair of contradictory theorems. With the means here at our disposal, however, it would be a hopeless undertaking to sketch the proof of the consistency or even to try to acquaint the reader with its fundamental idea; this would require a much deeper knowledge of logic, and an essential preliminary task would be the reconstruction of the part of arithmetic in question as a formalized deductive theory (cf. Section 40). It may be added that, if System 𝔄 is enriched by a single sentence to the effect that at least two distinct numbers exist, then the attempt to prove the consistency of the axiom system thus extended will meet with difficulties of the same degree as are encountered in the case of the entire system of arithmetic.

60. Problem of the completeness of the constructed theory

In comparison with the question of the consistency, that of the completeness of System 𝔄 can be dealt with much more readily.

There are numerous problems, formulated exclusively in logical terms and in primitive terms of System 𝔄, that do not in any way admit of a decision on the basis of this system. One such problem has already been mentioned in the preceding section. Another example is given by the sentence stating that to any number x, there exists a number y such that

$$x = y + y.$$

On the basis of the axioms of System 𝔄 alone, it is impossible either to prove or to disprove this sentence. That this is so can be seen from the following consideration. By the symbol "**N**" we have denoted the set of all real numbers; that is to say, the set **N** comprehends the integers as well as the fractions, the rational numbers as well as the irrational ones. But it can be seen at once that none of the axioms and, hence, none of theorems following from them would lose their validity if by the symbol "**N**" we were to denote either the set of all integers (the positive and negative ones including the number 0), or the set of all rational numbers; that is to say, all these statements would remain valid if the word "*number*" meant either "*integer*" or "*rational number*". In the first case, the sentence mentioned above, which states that for any given number there is another number half as large, would be false; in the second case it would be true. If, therefore, we succeeded in proving this sentence on the basis of System 𝔄, we would arrive at a contradiction within the arithmetic of integers; if, on the other hand, we were able to disprove it, we would find ourselves involved in a contradiction within the arithmetic of rational numbers.

The argument sketched just now falls under the category of proofs by interpretation (cf. Sections 37 and 56); in order to make this clear let us reformulate the argument slightly. Let "**I**" denote the set of all integers, and "**R**" the set of all rational numbers. We shall now give two interpretations of System 𝔄 within arithmetic. The symbols "$<$", "$>$" and "$+$" remain unchanged in both interpretations, while the symbol "**N**" which occurs explicitly or implictly in each of the axioms is to be replaced by "**I**" in the first and by "**R**" in the second interpretation. (We disregard here the remarks made in Section 43 concerning the possible elimination of the symbol "**N**", since this would slightly complicate our reasoning.) All axioms of System 𝔄 retain their validity in both interpretations; the sentence:

for every number x there exists a number y such that

$$x = y + y,$$

however, is fulfilled only in the case of the second interpretation, while in the case of the first interpretation its negation holds:

not for every number x is there a number y such that

$$x = y + y.$$

On the assumption of the consistency of arithmetic we conclude from the first interpretation that the sentence in question cannot be proved on the basis of System 𝔄, and from the second interpretation we conclude that it also cannot be disproved.

We have thus shown that there exist two contradictory sentences, formulated exclusively in logical terms and in primitive terms of the mathematical theory which we have been considering, with the property that neither of them can be derived from the axioms of that theory. Consequently we have:

The mathematical theory based on System $\mathfrak{A}$ is incomplete.

Exercises

1. Let us agree that the formula:

$$x \circledS y$$

means the same as:

$$x + 1 < y.$$

Now replace, in the axioms of System $\mathfrak{A}^{(2)}$ of Section 57, the symbol "$<$" throughout by "$\circledS$", and determine which of the axioms retain their validity and which do not, and hence infer that Axiom $1^{(2)}$ cannot be derived from the remaining axioms. What is the name of the method of inference here applied?

2. Following the lines of the independence proof sketched in Section 56 for Axiom $2^{(1)}$, show that Axiom $2^{(2)}$ cannot be derived from the remaining axioms of System $\mathfrak{A}^{(2)}$.

3. Let the symbol "$\mathring{\mathbf{N}}$" designate the set consisting of the three numbers 0, 1 and 2. Among the elements of this set we define a relation $\mathring{<}$, stipulating that it should hold only in these three cases:

$$0 \mathring{<} 1, \quad 1 \mathring{<} 2, \quad 2 \mathring{<} 0.$$

Further, we define the operation $\mathring{+}$ on the elements of the set $\mathring{\mathbf{N}}$ by the following formulas:

$$0 \mathring{+} 0 = 1 \mathring{+} 2 = 2 \mathring{+} 1 = 0,$$
$$0 \mathring{+} 1 = 1 \mathring{+} 0 = 2 \mathring{+} 2 = 1,$$
$$0 \mathring{+} 2 = 1 \mathring{+} 1 = 2 \mathring{+} 0 = 2.$$

Now replace, in the axioms of System $\mathfrak{A}^{(2)}$, the primitive terms of that system by "$\mathring{\mathbf{N}}$", "$\mathring{<}$" and "$\mathring{+}$" respectively (and the word "*number*" by the expression "*one of the three numbers* 0, 1 *and* 2"); show, by doing so, that Axiom $3^{(2)}$ cannot be derived from the remaining axioms.

4. In order to show by means of a proof by interpretation that Axiom $4^{(2)}$ is not derivable from the remaining axioms of System $\mathfrak{A}^{(2)}$, it is sufficient to replace the symbol of addition in all axioms by the symbol of a certain one among the four operations mentioned in Exercise 3 of Chapter VIII. Which is the operation that has to be used?

5. Consider the operation $\oplus$ satisfying the following formula:

$$x \oplus y = 2 \cdot (x + y).$$

Show, with the help of this operation, that Axiom $5^{(2)}$ cannot be deduced from the other axioms of System $\mathfrak{A}^{(2)}$.

6. Construct a set of numbers of such a kind that, together with the relation $<$ and the operation $+$, it fails to satisfy Axiom $6^{(2)}$ but forms a model of the remaining axioms of System $\mathfrak{A}^{(2)}$. What conclusion may, therefore, be drawn with respect to the possibility of deriving Axiom $6^{(2)}$?

7. In order to show that Axiom $7^{(2)}$ is not a consequence of the other axioms of System $\mathfrak{A}^{(2)}$, one can proceed by replacing in all axioms two of the primitive terms of the system by corresponding symbols introduced in Exercise 3, leaving the third primitive term unchanged. Determine which term should be left unchanged.

8. The results obtained in Exercises 1–7 go to show that none of the axioms of System $\mathfrak{A}^{(2)}$ can be derived from the remaining axioms of that system. Carry out analogous independence proofs for the axiom systems $\mathfrak{A}^{(1)}$ of Section 55 and $\mathfrak{A}^{(3)}$ of Section 58 (using, in part, the interpretations applied in the preceding exercises).

9. Show, on the basis of the axiom system $\mathfrak{A}^{(2)}$, that any set of numbers which is an Abelian group with respect to the operation of addition is at the same time an ordered Abelian group with respect to the relation *less than* and the operation of addition. Give examples of sets of numbers of this kind.

10. In Exercise 5 of Chapter VIII several sets of numbers were given which form Abelian groups with respect to multiplication. Which of these sets are ordered Abelian groups with respect to the relation *less than* and the operation of multiplication, and which are not?

11. Use the result obtained in Exercise 10 for a new proof of the independence of Axiom $7^{(2)}$ from the remaining axioms of System $\mathfrak{A}^{(2)}$ (cf. Exercise 7).

*12. On the basis of the axiom system $\mathfrak{A}^{(2)}$ prove the following theorem:

if there are at least two different numbers, then there is, for any number x, a number y such that $x < y$.

As a generalization of this result, prove the following general group-theoretical theorem:

if the class K is an ordered Abelian group with respect to the relation R and the operation O, and if K has at least two elements, then, for any element x of K, there exists an element y of K such that x R y.

Show with the help of this theorem that no class which is an ordered Abelian group can consist of exactly two, or three, and so on, elements. Can it consist of just one element? (Cf. Exercise 8 of Chapter VIII.)

*13. Show that the system of Axioms $1^{(2)}$–$3^{(2)}$ (of Section 57) is equipollent to the system consisting of Axiom $1^{(1)}$ and the following sentence:

if $x < y$, $y < z$, $z < t$, $t < u$ *and* $u < v$, *then* $v \nless x$.

As a generalization of this result establish the following general law of the theory of relations:

for the class K to be ordered by the relation R it is necessary and sufficient that R is connected in K and that it satisfies the following condition:

if x, y, z, t, u and v are any elements of K, and if $x\,R\,y$, $y\,R\,z$, $z\,R\,t$, $t\,R\,u$ *and* $u\,R\,v$, *then it is not the case that* $v\,R\,x$.

*14. Using the considerations of Sections 48, 55 and 58, show that the following three systems of sentences are equipollent:

(a) the system of Axioms 6–9 of Section 47;

(b) the system of Axioms $4^{(2)}$–$6^{(2)}$ of Section 57;

(c) the system of Axioms $3^{(3)}$ and $4^{(3)}$ of Section 58.

Generalizing this result, formulate new definitions of the expression:

the class K is an Abelian group with respect to the operation O,

that are equivalent to, but simpler than, the definition given in Section 47.

*15. Consider the system $\mathfrak{A}^{(4)}$ consisting of the following five axioms:

AXIOM $1^{(4)}$. *If* $x \neq y$, *then* $x < y$ *or* $y < x$.

AXIOM $2^{(4)}$. *If* $x < y$, $y < z$, $z < t$, $t < u$ *and* $u < v$, *then* $v \nless x$.

AXIOM $3^{(4)}$. $x + (y + z) = (x + z) + y$.

AXIOM $4^{(4)}$. *For any numbers* x *and* y *there exists a number* z *such that* $x = y + z$.

AXIOM $5^{(4)}$. *If* $y < z$, *then* $x + y < x + z$.

Using the results of Exercises 13 and 14, show that System $\mathfrak{A}^{(4)}$ is equipollent to each of the Systems $\mathfrak{A}^{(2)}$ and $\mathfrak{A}^{(3)}$.

16. In Section 58 it was asserted that the system of the three primitive terms "**N**", "$<$" and "$+$" is equipollent to the system of the terms "**N**", "$\leq$" and "$+$"; to this assertion it should really have been added that these systems are equipollent with respect to a certain system of sentences, for instance, with respect to System $\mathfrak{A}^{(3)}$ of Section 58 and Definition 1 of Section 46. Consider why such an addition is indispensable. In general, why is it necessary always to refer to a particular system of sentences when intending to establish the equipollence of two systems of terms (in the sense of Section 39)?

*17. Consider the system $\mathfrak{A}^{(5)}$ consisting of the following seven sentences:

AXIOM 1$^{(5)}$. *For any numbers x and y we have: $x \leqq y$ or $y \leqq x$.*

AXIOM 2$^{(5)}$. *If $x \leqq y$ and $y \leqq x$, then $x = y$.*

AXIOM 3$^{(5)}$. *If $x \leqq y$ and $y \leqq z$, then $x \leqq z$.*

AXIOM 4$^{(5)}$. $x + y = y + x$.

AXIOM 5$^{(5)}$. $x + (y + z) = (x + y) + z$.

AXIOM 6$^{(5)}$. *For any numbers x and y there exists a number z such that $x = y + z$.*

AXIOM 7$^{(5)}$. *If $y \leqq z$, then $x + y \leqq x + z$.*

Show that the axiom systems $\mathfrak{A}^{(2)}$ (of Section 57) and $\mathfrak{A}^{(5)}$ become equipollent systems of sentences, if Definition 1 of Section 46 is added to the first, and Theorem 8 of Section 46 to the second, considering the latter theorem as a definition of the symbol "$<$". Why may we not simply say that Systems $\mathfrak{A}^{(2)}$ and $\mathfrak{A}^{(5)}$ are equipollent?

18. Following the line of argument taken in Section 60, show that, on the basis of System $\mathfrak{A}$, the following sentence can be neither proved nor disproved:

if $x < z$, then there exists a number y for which $x < y$ and $y < z$.

*19. Show that, on the basis of System $\mathfrak{A}$, the following sentence can be neither proved nor disproved:

for any number x there exists a number y such that $x < y$.

*20. In the present chapter we have used the method of proof by interpretation in order to establish the independence or incompleteness of an axiom system. The same method is also employed in investigations concerning its consistency. In fact we have the following methodological law at our disposal which represents a consequence of the law of deduction:

If the deductive theory $\mathfrak{S}$ has an interpretation in the deductive theory $\mathfrak{T}$ and the theory $\mathfrak{T}$ is consistent, then the theory $\mathfrak{S}$ is also consistent.

Show that this statement is correct. In Section 38 some remarks have been made concerning possible interpretations of arithmetic and geometry; applying the law just given, deduce from these remarks consequences concerning the consistency of arithmetic and geometry and its connection with the consistency of logic.

· X ·

EXTENSION OF THE CONSTRUCTED THEORY

FOUNDATIONS OF ARITHMETIC OF REAL NUMBERS

61. First axiom system for the arithmetic of real numbers

The axiom system $\mathfrak{A}$ is insufficient as a foundation for the whole of the arithmetic of real numbers, because—as has been seen in Section 60—numerous theorems of this discipline cannot be deduced from the axioms of this system, and also for another, no less important and, incidentally, quite analogous reason: a number of concepts belonging to the field of arithmetic can be found that are not definable with the help of the primitive terms occurring in System $\mathfrak{A}$. Thus, System $\mathfrak{A}$ does not enable us to define the symbols of multiplication or division, or even such symbols as "1", "2", and so on.

The question at once presents itself as to how we have to transform and supplement our system of axioms and primitive terms in order to arrive at a sufficient basis for the construction of the entire arithmetic of real numbers. This problem can be solved in a variety of ways. Two different methods of solution will be sketched here.[1]

In the case of the first method, we choose as our point of departure the system $\mathfrak{A}^{(3)}$ (cf. Section 58); to the primitive terms appearing in that system we add the word "*one*" which, as usual,

[1] The first axiom system for the entire arithmetic of real numbers was published by Hilbert in 1900; this system is related to System $\mathfrak{A}''$ with which we shall become acquainted further below. Before the year 1900, axiom systems for certain less comprehensive parts of arithmetic had been known; the first system of this kind relating to the arithmetic of natural numbers was given in 1889 by Peano (cf. footnote 1 on p. 120). Several axiom systems for arithmetic and various parts of it—and, in particular, the first axiom system for the arithmetic of complex numbers—were published by Huntington (cf. footnote 8 on p. 140).

will be replaced by the symbol "1", and the axioms of the system are supplemented by four new sentences. In this way a new SYSTEM $\mathfrak{A}'$ is obtained, containing the four primitive terms "**N**", "$<$", "$+$" and "1" and consisting of the nine axioms which we shall list explicitly below:

AXIOM 1′. *If* $x \neq y$, *then* $x < y$ *or* $y < x$.

AXIOM 2′. *If* $x < y$, *then* $y \nless x$.

AXIOM 3′. *If* $x < z$, *then there exists a number* y *such that* $x < y$ *and* $y < z$.

AXIOM 4′. *If* K *and* L *are any sets of numbers* (i.e., $K \subset \mathbf{N}$ *and* $L \subset \mathbf{N}$) *satisfying the condition*:

for any x *belonging to* K *and any* y *belonging to* L, *we have*: $x < y$,

then there exists a number z *for which the following condition holds:*

if x *is any element of* K *and* y *any element of* L, *and if* $x \neq z$ *and* $y \neq z$, *then* $x < z$ *and* $z < y$.

AXIOM 5′. $x + (y + z) = (x + z) + y$.

AXIOM 6′. *For any numbers* x *and* y *there exists a number* z *such that* $x = y + z$.

AXIOM 7′. *If* $x + z < y + t$, *then* $x < y$ *or* $z < t$.

AXIOM 8′. $1 \in \mathbf{N}$.

AXIOM 9′. $1 < 1 + 1$.

62. Closer characterization of the first axiom system; its methodological advantages and didactical disadvantages

The axioms listed in the preceding section fall into three groups. In the first group, consisting of Axioms 1′–4′, only the two primitive terms "**N**" and "$<$" occur; in the second group, to which Axioms 5′–7′ belong, we have the additional symbol "$+$"; finally, in the third group, in which we have Axioms 8′ and 9′, the new symbol "1" appears.

Among the axioms of the first group there are two which we had not met before, namely Axioms 3′ and 4′. Axiom 3′ is called the LAW OF DENSITY for the relation *less than*—it expresses the fact that this relation is dense in the set of all numbers. In general we say that the relation R is DENSE IN THE CLASS K if, for any two elements x and y of this class, the formula:

$$x\,R\,y$$

always implies the existence of an element z of the class K for which

$$x\,R\,z \quad \text{and} \quad z\,R\,y$$

hold. Axiom 4′ is known as the LAW OF CONTINUITY for the relation *less than* or as the AXIOM OF CONTINUITY or, also, as DEDEKIND'S AXIOM[2]; in order to state in general, under what condition the relation R is called CONTINUOUS IN THE CLASS K, it is sufficient to replace, in Axiom 4′, "**N**" by "K" (and, in connection with that, the word "*number*" by the expression "*element of the class K*"), and further "$<$" by "R". If, in particular, the class K is ordered by the relation R, and if R is dense or continuous in K, then K is said to be DENSELY ORDERED or CONTINUOUSLY ORDERED, respectively.

Axiom 4′ is intuitively less evident and more complicated than the remaining axioms; for one thing, it differs from the other axioms inasmuch as it is concerned, not with individual numbers, but with sets of numbers. In order to give this axiom a simpler and more comprehensible form, it is expedient to have it preceded by the following definitions:

We say that the set of numbers K PRECEDES *the set of numbers L if, and only if, every number of K is less than every number of L*

We say that the number z SEPARATES *the sets of numbers K and L if, and only if, for any two elements x of K and y of L, both distinct from z, we have:* $x < z$ *and* $z < y$.

[2] This axiom—in a slightly more complicated formulation—originates with the German mathematician R. DEDEKIND (1831–1916), whose researches have contributed greatly to the foundations of arithmetic and, especially, of the theory of irrational numbers.

On the basis of these definitions we can give the axiom of continuity the following very simple formulation:

If one set of numbers precedes another, then there exists at least one number separating the two sets.

All the axioms of the second group are already known to us from earlier considerations. The axioms of the third group, though new, have so simple and obvious a content that they hardly require any comment. We might only remark this much, that if Axiom 9′ is preceded by definitions of the symbol "0" and of the expression "*positive number*", then it may be replaced either by the formula:

$$0 < 1$$

or else by the sentence:

1 *is a positive number.*

The Axioms 1′, 2′, 5′, 6′ and 7′ form just what we called System $\mathfrak{A}^{(3)}$ which—like the equipollent System $\mathfrak{A}^{(2)}$—characterizes the set of all numbers as an ordered Abelian group (cf. Section 57). Considering the content of the newly added Axioms 3′, 4′, 8′ and 9′, we may now describe the whole system as follows:

System $\mathfrak{A}'$ expresses the fact that the set of all numbers is a densely and continuously ordered Abelian group with respect to the relation $<$ and the operation of addition, and it singles out a certain positive element 1 *in that set.*

From the methodological point of view, System $\mathfrak{A}'$ possesses several advantages. Formally considered, it appears to be the simplest of all known axiom systems that form a sufficient basis upon which to found the entire system of arithmetic. With the exception of Axiom 1′, which—though not quite easily—can be derived from the remaining axioms, all the other axioms of the system as well as the primitive terms occurring in these axioms are mutually independent. The didactical value of System $\mathfrak{A}'$, on the other hand, is far smaller, because the simplicity of the foundations causes considerable complications in the further construction. Even the definition of multiplication and the derivation of the basic laws for this operation are not easy to carry through. Almost from the very beginning, the arguments will have to make essential use of the continuity axiom (without its help, for instance, it would not be possible to prove, on the basis of System $\mathfrak{A}'$, the existence of the number $\frac{1}{2}$, i.e. of a number y such that $y + y = 1$), and the inferences based on that axiom are usually found rather difficult by the beginner.

63. Second axiom system for the arithmetic of real numbers

For the reasons mentioned above it is worth while to search for a different axiom system upon which to construct arithmetic. A system of this kind can be obtained in the following way. As our point of departure we use System $\mathfrak{A}^{(2)}$. Three new primitive terms will be adopted: "*zero*", "*one*" and "*product*"; the first two will, as usual, be replaced by the symbols "0" and "1", while, instead of the expression "*the product of the numbers* (or *factors*) *x and y*" (or "*the result of multiplying x and y*") we shall use the customary symbol "$x \cdot y$". Further, thirteen new axioms will be added; of these, two are already known to us, namely the axiom of continuity and the law of performability for addition. We thus finally arrive at SYSTEM $\mathfrak{A}''$ containing the six primitive terms "**N**", "$<$", "$+$", "0", "$\cdot$" and "1", and consisting of the following twenty sentences:

AXIOM 1″. *If* $x \neq y$, *then* $x < y$ *or* $y < x$.

AXIOM 2″. *If* $x < y$, *then* $y \nless x$.

AXIOM 3″. *If* $x < y$ *and* $y < z$, *then* $x < z$.

AXIOM 4″. *If K and L are any sets of numbers satisfying the condition:*

for any x belonging to K and any y belonging to L, we have: $x < y$,

then there exists a number z for which the following condition holds:

if x is any element of K and y any element of L, and if $x \neq z$ *and* $y \neq z$, *then* $x < z$ *and* $z < y$.

AXIOM 5″. *For any numbers y and z there exists a number x such that* $x = y + z$ (in other words: *if* $y \in \mathbf{N}$ *and* $z \in \mathbf{N}$, *then* $y + z \in \mathbf{N}$).

AXIOM 6″. $x + y = y + x$.

AXIOM 7″. $x + (y + z) = (x + y) + z$.

AXIOM 8″. *For any numbers x and y there exists a number z such that* $x = y + z$.

AXIOM 9″. *If* $y < z$, *then* $x + y < x + z$.

AXIOM 10″. $0 \in \mathbf{N}$.

AXIOM 11″. $x + 0 = x$.

AXIOM 12″. *For any numbers y and z there exists a number x such that* $x = y \cdot z$ (in other words: *if* $y \in \mathbf{N}$ *and* $z \in \mathbf{N}$, *then* $y \cdot z \in \mathbf{N}$).

AXIOM 13″. $x \cdot y = y \cdot x$.

AXIOM 14″. $x \cdot (y \cdot z) = (x \cdot y) \cdot z$.

AXIOM 15″. *For any numbers x and y, if* $y \neq 0$, *there exists a number z such that* $x = y \cdot z$.

AXIOM 16″. *If* $0 < x$ *and* $y < z$, *then* $x \cdot y < x \cdot z$.

AXIOM 17″. $x \cdot (y + z) = (x \cdot y) + (x \cdot z)$.

AXIOM 18″. $1 \in \mathbf{N}$.

AXIOM 19″. $x \cdot 1 = x$.

AXIOM 20″. $0 \neq 1$.

64. Closer characterization of the second axiom system; concepts of a field and of an ordered field

In System $\mathfrak{A}''$, as in System $\mathfrak{A}'$, three groups of axioms may be distinguished. In Axioms 1″–4″, which form the first group, we have only the two primitive terms "**N**" and "$<$"; the second group, consisting of Axioms 5″–11″, contains the two further symbols: the addition sign "+" and the symbol "0"; finally, the third group, which is made up of Axioms 12″–20″, involves primarily the multiplication sign "·" and the symbol "1".

All axioms of the first two groups, with the exception of Axioms 10″ and 11″, are already known to us. Axioms 10″ and 11″ together state that 0 is a (right-hand) unit element of the operation of addition. For, in general, an element u is said to be a RIGHT-HAND OR LEFT-HAND UNIT ELEMENT OF THE OPERATION O IN THE CLASS K, if u belongs to K and if every element x of K satisfies the formula:

$$x \, O \, u = x, \quad \text{or} \quad u \, O \, x = x,$$

respectively. If u is both a right- and left-hand unit element it is simply called a UNIT ELEMENT OF THE OPERATION O IN THE CLASS K; evidently, in the case of a commutative operation O, every right- or left-hand unit element is simply a unit element.

In the first three axioms of the third group, i.e. Axioms 12″–14″, we recognize the LAW OF PERFORMABILITY and the COMMUTATIVE and ASSOCIATIVE LAWS for multiplication; they correspond precisely to Axioms 5″–7″. Axioms 15″ and 16″ are called the LAW OF RIGHT INVERTIBILITY for multiplication and the LAW OF MONOTONY for multiplication with respect to the relation *less than.* These axioms correspond to the laws of invertibility and monotony for addition, but not quite exactly. The difference lies in the fact that their hypotheses contain the restrictive conditions "$y \neq 0$" and "$0 < x$"; in spite of their names, therefore, they do not permit us to assert simply that multiplication is invertible, or that it is monotonic with respect to the relation $<$ (in the sense of Sections 47 and 49).

Axiom 17″ establishes a fundamental connection between addition and multiplication; it is the so-called DISTRIBUTIVE LAW (or, strictly speaking, the LAW OF RIGHT DISTRIBUTIVITY) for multiplication with respect to addition. In general, the operation P is called RIGHT- OR LEFT-DISTRIBUTIVE WITH RESPECT TO THE OPERATION O IN THE CLASS K if any three elements x, y and z of the class K satisfy the formula:

$$x \, P \, (y \, O \, z) = (x \, P \, y) \, O \, (x \, P \, z),$$

$$\text{or} \quad (x \, O \, y) \, P \, z = (x \, P \, z) \, O \, (y \, P \, z),$$

respectively. If the operation P is commutative, the notions of right and left distributivity coincide, and we simply say that the operation P is DISTRIBUTIVE WITH RESPECT TO THE OPERATION O IN THE CLASS K.

The last three axioms concern the number 1. Axioms 18″ and 19″ together state that 1 is a right-hand unit element of the operation of multiplication. The content of Axiom 20″ does not call for any explanation; the role played by this axiom in the construction of arithmetic is greater than might at first be supposed, for without its help it is impossible to show that the set of all numbers is infinite.

In order to describe briefly the totality of properties attributed to addition and multiplication in Axioms 5″–8″, 12″–15″ and 17″, one says that these axioms characterize the set **N** as a FIELD (or, more precisely, a COMMUTATIVE FIELD) WITH RESPECT TO THE OPERATIONS OF ADDITION AND MULTIPLICATION. If, in addition, the axioms of order 1″–3″ and the axioms of monotony 9″ and 16″ are taken into account, the set **N** is said to be characterized as an ORDERED FIELD WITH RESPECT TO THE RELATION $<$ AND THE OPERATIONS OF ADDITION AND MULTIPLICATION. The reader will easily guess how the concepts of a field and of an ordered field are to be extended to arbitrary classes, operations and relations.—If, finally, the continuity axiom 4″ and the axioms concerning the numbers 0 and 1, i.e. Axioms 10″, 11″, 18″–20″, are taken into consideration, then the content of the entire axiom system $\mathfrak{A}''$ may be described as follows:

System $\mathfrak{A}''$ *expresses the fact that the set of all numbers is a continuously ordered field with respect to the relation* $<$ *and the operations of addition and multiplication, and singles out two distinct elements* 0 *and* 1 *in that set, of which the first is the unit element of addition and the second the unit element of multiplication.*

65. Equipollence of the two axiom systems; methodological disadvantages and didactical advantages of the second system

The axiom systems $\mathfrak{A}'$ and $\mathfrak{A}''$ are equipollent (or, rather, they become equipollent as soon as the first system is supplemented by the definitions of the symbol "0" and of the multiplication sign "·", which can be formulated with the help of its primitive terms). However, the proof of this equipollence is not easy. It is true that the derivation of the axioms of the first system from those of the second is not especially difficult; but as far as the opposite task is concerned, it already follows from our earlier remarks that, on the basis of the first system, both the definition of multiplication and the proof of the basic laws governing this operation (which occur as axioms in the second system) present considerable difficulties.

In methodological respects System 𝔄′ surpasses System 𝔄″ considerably. The number of axioms in 𝔄″ is more than twice as large. The axioms are not mutually independent; thus, for instance, Axioms 5″ and 12″, i.e. the laws of performability for addition and multiplication, are derivable from the remaining axioms, or, if these two axioms are retained, certain others such as Axioms 6″, 11″ and 14″ may be eliminated. The primitive terms, too, are not independent, for three of them, namely "<", "0" and "1", can be defined in terms of the others (*one of the possible definitions of the symbol "0" has been stated in Section 53*), and consequently the number of axioms can be further reduced.

We see, therefore, that System 𝔄″ admits of important simplifications of various kinds; but as a consequence of these simplifications the didactical advantages of the system would be diminished considerably. And these advantages are indeed great. On the basis of System 𝔄″ it is possible to develop without any difficulty the most important parts of the arithmetic of real numbers,—such as the theory of the fundamental relations among numbers, the theory of the four elementary arithmetical operations of addition, subtraction, multiplication and division, the theory of linear equations, inequalities and functions. The methods of inference to be applied here are of a very natural and quite elementary character; in particular, the axiom of continuity does not enter at all at this stage, it plays an essential role only when we go over to the "higher" arithmetical operations of raising to a power, of extracting roots and of taking logarithms, and it is indispensable for the proof of the existence of irrational numbers. No other system of axioms and primitive terms appears to be known that might furnish a more advantageous basis for an elementary and, at the same time, strictly deductive construction of the arithmetic of real numbers.

Exercises

1. Show that the set of all positive numbers, the relation <, the operation of multiplication and the number 2 form a model of System $\mathfrak{A}'$ and that, therefore, this system possesses at least two different interpretations within arithmetic.

2. Which of the relations listed in Exercise 6 of Chapter V are dense?

*3. How can we express—in the symbolism of the calculus of relations—the fact that the relation R is dense (in the universal class)? How can we express the fact that the relation R is both transitive and dense by means of one equation from the calculus of relations? (Cf. Exercise 17 of Chapter V.)

4. Which of the following sets of numbers are densely ordered by the relation <:

(a) the set of all natural numbers,

(b) the set of all integers,

(c) the set of all rational numbers,

(d) the set of all positive numbers,

(e) the set of all numbers different from 0?

*5. In order to prove, on the basis of system $\mathfrak{A}'$, the existence of the number $\frac{1}{2}$, i.e. of a number z such that:

$$z + z = 1,$$

we can proceed as follows. Let K be the set of all numbers x such that:

$$x + x < 1,$$

and, similarly, let L be the set of all numbers y such that:

$$1 < y + y.$$

We show first that the set K precedes the set L. Applying now the axiom of continuity, we obtain a number z separating the sets K and L. Next it can be shown that the number z can belong neither to K (otherwise a number x in K greater than z would exist) nor to L. From this we can conclude that z is the number looked for, in other words, that

$$z + z = 1$$

holds. Carry out in detail the proof sketched above.

*6. Generalizing the procedure of the preceding exercise, prove the following theorem T on the basis of System $\mathfrak{A}'$:

T. *For any number x there exists a number y such that* $x = y + y$.

Compare the result obtained in this way with the remarks in Section 60.

*7. Replace, in System $\mathfrak{A}'$, Axiom 3′ by Theorem T of the previous exercise. Show that the system of sentences obtained thereby is equipollent to System $\mathfrak{A}'$.

Hint: In order to derive Axiom 3′ from the modified system, substitute "$x + z$" for "x" in T; in view of the hypothesis of Axiom 3′, it can then be shown easily that the number y fulfils the conclusion of this axiom.

*8. Use the method of proof by interpretation to show that, after omission of Axiom 1, System $\mathfrak{A}'$ becomes a system of mutually independent axioms.

*9. Give a geometric interpretation of the axiom systems $\mathfrak{A}'$ and $\mathfrak{A}''$, by way of an extension of Exercise 2 of Chapter VIII.

*10. Write all axioms of Systems $\mathfrak{A}'$ and $\mathfrak{A}''$ in logical symbolism.

11. Do the operations of subtraction and division and the operations mentioned in Exercise 3 of Chapter VIII possess right- or left-hand unit elements, or simply unit elements in the set of all numbers? Do the operations of addition and multiplication of point sets possess unit elements in the class of all point sets?

*12. Show that any operation commutative in a class possesses at most one unit element in that class. Further, in generalization of the result obtained in Exercise 24 of Chapter VIII, prove the following group-theoretical theorem:

if the class K is an Abelian group with respect to the relation O, then the operation O possesses exactly one unit element in the class K.

13. Consider the five arithmetical operations of addition, subtraction, multiplication, division and raising to a power. Formulate the sentences asserting that one of these operations is right- and left-distributive with respect to another (there are altogether 40 such sentences), and determine which of these are true.

14. Solve the same problem as in the preceding exercise with respect to the four operations **A, B, G** and **L** introduced in Exercise 3 of Chapter VIII. Further, show that every operation performable in a certain set of numbers is right- and left-distributive in that set with respect to the operations **A** and **B**.

15. Is the addition of classes distributive with respect to multiplication, and vice versa? (Cf. Exercise 15 of Chapter IV.)

16. Which of the sets of numbers listed in Exercise 4 are fields with respect to addition and multiplication or ordered fields with respect to these operations and the relation $<$?

17. Show that the set consisting of the numbers 0 and 1 is a field with respect to the operation $\oplus$ defined in Exercise 6 of Chapter VIII and to multiplication.

18. Find two operations on the numbers 0, 1 and 2, of such a kind that the set of these three numbers forms a field with respect to those two operations.

19. How is it possible to define the symbol "1" with the help of multiplication?

20. The following theorem can be derived from the axioms of System $\mathfrak{A}''$:

if $0 < x$, *then there exists a number* y *such that* $x = y \cdot y$.

Supposing this theorem to have been proved already, derive with its help from the axioms of System $\mathfrak{A}''$ the following theorem:

$x < y$ *if, and only if,* $x \neq y$ *and if there exists a number* z *such that* $x + z \cdot z = y$.

Does this theorem justify a remark made in Section 65 concerning a possible reduction in the number of primitive terms of System $\mathfrak{A}''$?

*21. Prove Theorem T of Exercise 6 on the basis of System $\mathfrak{A}''$. Compare this proof with the one suggested in Exercise 6 with reference to System $\mathfrak{A}'$; which of the two proofs is more difficult and requires a greater knowledge of logical concepts?

Hint: In order to derive Theorem T from System $\mathfrak{A}''$, apply Axiom 15″, with "y" and "z" replaced by "$1 + 1$" and "y" respectively (it has, however, first to be shown that $1 + 1$ is distinct from 0); thereby a number y is obtained, of which it can be shown with the help of Axioms 13″, 17″ and 19″ that it fulfils the formula given in Theorem T.

*22. Derive all the axioms of System $\mathfrak{A}'$ from the axioms of System $\mathfrak{A}''$.

Hint: In order to deduce Axiom 3′, assume Theorem T of Exercise 6 to have already been proved on the basis of System $\mathfrak{A}''$ (cf. the previous exercise); from there on proceed in the same way as in Exercise 7.

SUGGESTED READINGS

In concluding this book we should like to point out to the reader a number of works which may be of service to him in deepening and extending the knowledge acquired here. However, none of the works listed below offers a systematic and exhaustive treatment of all the problems upon which we have touched. As it is, the literature of the field in which we are interested is as yet comparatively poorly supplied with textbooks, and it is hard to name many books whose presentation combines comprehensibility with the required degree of exactitude.

E. V. Huntington. *The Fundamental Propositions of Algebra.* Monograph IV in *Monographs on Topics of Modern Mathematics relevant to the Elementary Field*, edited by J. W. A. Young. New York 1911.

We commend this work to the attention of those readers who are interested in the considerations contained in the last two chapters of the present book. They will find there a simple, precise and clear presentation of the results of methodological investigations into the axiomatic foundations of the arithmetic of real and complex numbers.

C. I. Lewis. *A Survey of Symbolic Logic.* Berkeley 1918.

Though the systematic part of this book is somewhat outdated, its historical part can be warmly recommended, since it offers a great deal of interesting and instructive information on the development of modern logic.

C. I. Lewis and C. H. Langford. *Symbolic Logic.* New York 1932.

This book contains much interesting and relevant material from various parts of symbolic logic and the methodology of deductive sciences, and, especially, from the domain of sentential calculus and its methodology. It cannot, however, serve as a systematic text of logic and methodology, since it fails to touch on many important topics belonging to these fields. The valuable historical material contained in Professor Lewis's work mentioned above has not been included in this book.

B. Russell. *Introduction to Mathematical Philosophy.* London 1921 (2d edition).

This work gives a clear and easily intelligible presentation of the most important concepts of modern logic, especially of those necessary for the establishment of mathematics as a part of logic. It treats among other things of many topics not discussed or only superficially touched upon in the present book, such as the theory of types or the problems connected with the axiom of infinity and the multiplicative axiom; it can serve as a preparatory text for the study of the work *Principia Mathematica* listed below.

J. W. YOUNG. *Lectures on Fundamental Concepts of Algebra and Geometry.* New York 1911.

In this extremely informative little book the reader will find many interesting considerations and examples from the domain of the methodology of mathematics. Moreover, the reader may acquaint himself here with some of the basic concepts of the general theory of sets.

J. H. WOODGER. *The Technique of Theory Construction.* Chicago 1939. (International Encyclopedia of Unified Science, vol. 2, no. 5.)

This short monograph will acquaint the reader by means of a concrete example with the technique of constructing formalized deductive theories; it contains also a discussion of general methodological problems and interesting considerations as to the possibility and usefulness of applying deductive methods within empirical sciences. It can be warmly recommended, especially to readers interested in the last-mentioned problem.

The following two works are much more difficult:

*R. CARNAP. *The Logical Syntax of Language.* New York and London 1937.

This book is interesting but not easy. As for its content, it corresponds to what we have called the methodology of deductive sciences, but in that wider conception which was discussed in Section 42. Emphasis is laid, however, not so much on a presentation of the results achieved in this field, as on the development of the conceptual apparatus. The systematic, deductive part of the book is treated rather sketchily, and it calls for quite a considerable amount of skill in abstract deductive thinking on the part of the reader, to enable him to fill the gaps which he will find and even, in places, to introduce some corrections in the arguments of the author. The book should be recommended, first of all, to those readers who are interested in the question of the significance of methodological investigations for general philosophical problems; they will find numerous remarks on this subject throughout the text and, moreover, more systematized considerations on it in the last part of the book.

*A. N. Whitehead and B. Russell. *Principia Mathematica.* Vol. 1–3. Cambridge 1925 and 1927 (2d edition).

This work has already been quoted several times in the present book. It is undoubtedly the most representative work of modern logic, and as for the influence it has exerted it has been no less than epoch-making in the development of logical investigations. The purpose which the authors had in mind was to construct a complete system of logic which would provide a sufficient basis for the foundations of mathematics. This task was fulfilled in a very thorough and exhaustive manner (though not in every detail complying with the very strictest requirements of present-day methodology). By most people the study of this work, which is written preponderantly in symbolic language, will not be found easy, but it is indispensable for anybody who desires to acquire a thorough knowledge of the conceptual apparatus of modern logic.

From the foreign literature we recommend a few books, exact equivalents of which are not available in English:

R. Carnap. *Abriss der Logistik.* Vienna 1929. (Schriften zur wissenschaftlichen Weltauffassung, vol. 2.)

In spite of its shortness this book gives a clear survey of all important concepts of contemporary logic, thereby constituting a popularization of the above-mentioned work *Principia Mathematica*; moreover, we find there interesting examples of applying logical concepts and methods in other sciences.

K. Grelling. *Mengenlehre.* Leipzig and Berlin 1924. (Mathematisch-physikalische Bibliothek, vol. 58.)

We recommend this easily intelligible little book to those readers who desire to become acquainted with the most important concepts and fundamental results of the general theory of sets,—without intending to devote themselves to more profound studies in this field.

D. Hilbert and W. Ackermann. *Grundzüge der theoretischen Logik.* Berlin 1938 (2d edition). (Grundlehren der mathematischen Wissenschaften, vol. 27.)

This book outlines the more elementary and fundamental parts of mathematical logic and discusses the most important methodological problems concerning these parts of logic. It is both precise and intelligible, and may well serve as an introduction to a systematic study and more profound investigations in the fields of logic and methodology.

H. Scholz. *Geschichte der Logik.* Berlin 1931. (Geschichte der Philosophie in Längsschnitten, vol. 4.)

We have here a historical survey of the development of logic up to the present time. The belief in the far-reaching role of contemporary mathematical logic permeates every page of this book, and the vivacity and colourfulness of the style will make the book attractive reading even to a layman.

As a more advanced work we list the following:

*D. Hilbert and P. Bernays. *Grundlagen der Mathematik.* Vol. 1–2. Berlin 1934 and 1939. (Grundlehren der mathematischen Wissenschaften, vol. 40 and 50.)

This book is a very comprehensive, though not exhaustive, exposition of the more important results obtained so far in the methodology of mathematics (in that conception of this science which was discussed in Section 42). In accordance with the theoretical outlook of the authors in the question of methodological investigations, emphasis is laid on the results obtained by means of so-called constructive methods. Without a thorough study of this work, which is unique in its kind in the literature, it would scarcely be possible to conduct independent research work in the field of methodology. The reader may find some difficulties in studying the book, and the reasons for the chosen arrangement of the material may perhaps not always seem obvious to him; but shortcomings of this kind are, after all, hardly avoidable, in view of the fact that the results presented in the work are quite recent and belong to a field which is still in the process of intensive development.

In conclusion we may mention that there exists in the United States a special society, the *Association for Symbolic Logic,* which brings together all scientific workers in the fields of logic and methodology. Since 1936, the Association has published its own quarterly, the *Journal of Symbolic Logic,* edited by A. Church and E. Nagel, in which both original contributions and reviews of the entire current logical literature are published. Volume 1 (1936) of this periodical contains an exhaustive bibliography, assembled by Church, of all publications in the domain of mathematical logic during the whole period of its existence up to the year 1935; a supplement to this bibliography appeared in Volume 3 (1938).

INDEX

The index contains the more important terms and expressions used in this book in a technical meaning as well as the names of scientists mentioned in the text. The square brackets following a term enclose expressions used in the book as synonyms of that term.

All numbers refer to pages; numbers in bold type indicate the main references and most important passages.

A CATALOG OF SELECTED
DOVER BOOKS
IN SCIENCE AND MATHEMATICS

Astronomy

BURNHAM'S CELESTIAL HANDBOOK, Robert Burnham, Jr. Thorough guide to the stars beyond our solar system. Exhaustive treatment. Alphabetical by constellation: Andromeda to Cetus in Vol. 1; Chamaeleon to Orion in Vol. 2; and Pavo to Vulpecula in Vol. 3. Hundreds of illustrations. Index in Vol. 3. 2,000pp. 6¹/₈ x 9¹/₄.
Vol. I: 0-486-23567-X
Vol. II: 0-486-23568-8
Vol. III: 0-486-23673-0

EXPLORING THE MOON THROUGH BINOCULARS AND SMALL TELESCOPES, Ernest H. Cherrington, Jr. Informative, profusely illustrated guide to locating and identifying craters, rills, seas, mountains, other lunar features. Newly revised and updated with special section of new photos. Over 100 photos and diagrams. 240pp. 8¹/₄ x 11. 0-486-24491-1

THE EXTRATERRESTRIAL LIFE DEBATE, 1750–1900, Michael J. Crowe. First detailed, scholarly study in English of the many ideas that developed from 1750 to 1900 regarding the existence of intelligent extraterrestrial life. Examines ideas of Kant, Herschel, Voltaire, Percival Lowell, many other scientists and thinkers. 16 illustrations. 704pp. 5³/₈ x 8¹/₂. 0-486-40675-X

THEORIES OF THE WORLD FROM ANTIQUITY TO THE COPERNICAN REVOLUTION, Michael J. Crowe. Newly revised edition of an accessible, enlightening book re-creates the change from an earth-centered to a sun-centered conception of the solar system. 242pp. 5³/₈ x 8¹/₂. 0-486-41444-2

ARISTARCHUS OF SAMOS: The Ancient Copernicus, Sir Thomas Heath. Heath's history of astronomy ranges from Homer and Hesiod to Aristarchus and includes quotes from numerous thinkers, compilers, and scholasticists from Thales and Anaximander through Pythagoras, Plato, Aristotle, and Heraclides. 34 figures. 448pp. 5³/₈ x 8¹/₂.
0-486-43886-4

A COMPLETE MANUAL OF AMATEUR ASTRONOMY: TOOLS AND TECHNIQUES FOR ASTRONOMICAL OBSERVATIONS, P. Clay Sherrod with Thomas L. Koed. Concise, highly readable book discusses: selecting, setting up and maintaining a telescope; amateur studies of the sun; lunar topography and occultations; observations of Mars, Jupiter, Saturn, the minor planets and the stars; an introduction to photoelectric photometry; more. 1981 ed. 124 figures. 25 halftones. 37 tables. 335pp. 6¹/₂ x 9¹/₄. 0-486-42820-8

AMATEUR ASTRONOMER'S HANDBOOK, J. B. Sidgwick. Timeless, comprehensive coverage of telescopes, mirrors, lenses, mountings, telescope drives, micrometers, spectroscopes, more. 189 illustrations. 576pp. 5⁵/₈ x 8¹/₄. (Available in U.S. only.)
0-486-24034-7

STAR LORE: Myths, Legends, and Facts, William Tyler Olcott. Captivating retellings of the origins and histories of ancient star groups include Pegasus, Ursa Major, Pleiades, signs of the zodiac, and other constellations. "Classic."—Sky & Telescope. 58 illustrations. 544pp. 5³/₈ x 8¹/₂. 0-486-43581-4

Chemistry

THE SCEPTICAL CHYMIST: THE CLASSIC 1661 TEXT, Robert Boyle. Boyle defines the term "element," asserting that all natural phenomena can be explained by the motion and organization of primary particles. 1911 ed. viii+232pp. 5³/₈ x 8¹/₂. 0-486-42825-7

RADIOACTIVE SUBSTANCES, Marie Curie. Here is the celebrated scientist's doctoral thesis, the prelude to her receipt of the 1903 Nobel Prize. Curie discusses establishing atomic character of radioactivity found in compounds of uranium and thorium; extraction from pitchblende of polonium and radium; isolation of pure radium chloride; determination of atomic weight of radium; plus electric, photographic, luminous, heat, color effects of radioactivity. ii+94pp. 5³/₈ x 8¹/₂. 0-486-42550-9

CHEMICAL MAGIC, Leonard A. Ford. Second Edition, Revised by E. Winston Grundmeier. Over 100 unusual stunts demonstrating cold fire, dust explosions, much more. Text explains scientific principles and stresses safety precautions. 128pp. 5³/₈ x 8¹/₂. 0-486-67628-5

MOLECULAR THEORY OF CAPILLARITY, J. S. Rowlinson and B. Widom. History of surface phenomena offers critical and detailed examination and assessment of modern theories, focusing on statistical mechanics and application of results in mean-field approximation to model systems. 1989 edition. 352pp. 5³/₈ x 8¹/₂. 0-486-42544-4

CHEMICAL AND CATALYTIC REACTION ENGINEERING, James J. Carberry. Designed to offer background for managing chemical reactions, this text examines behavior of chemical reactions and reactors; fluid-fluid and fluid-solid reaction systems; heterogeneous catalysis and catalytic kinetics; more. 1976 edition. 672pp. 6¹/₈ x 9¹/₄. 0-486-41736-0 $31.95

ELEMENTS OF CHEMISTRY, Antoine Lavoisier. Monumental classic by founder of modern chemistry in remarkable reprint of rare 1790 Kerr translation. A must for every student of chemistry or the history of science. 539pp. 5³/₈ x 8¹/₂. 0-486-64624-6

MOLECULES AND RADIATION: An Introduction to Modern Molecular Spectroscopy. Second Edition, Jeffrey I. Steinfeld. This unified treatment introduces upper-level undergraduates and graduate students to the concepts and the methods of molecular spectroscopy and applications to quantum electronics, lasers, and related optical phenomena. 1985 edition. 512pp. 5³/₈ x 8¹/₂. 0-486-44152-0

A SHORT HISTORY OF CHEMISTRY, J. R. Partington. Classic exposition explores origins of chemistry, alchemy, early medical chemistry, nature of atmosphere, theory of valency, laws and structure of atomic theory, much more. 428pp. 5³/₈ x 8¹/₂. (Available in U.S. only.) 0-486-65977-1

GENERAL CHEMISTRY, Linus Pauling. Revised 3rd edition of classic first-year text by Nobel laureate. Atomic and molecular structure, quantum mechanics, statistical mechanics, thermodynamics correlated with descriptive chemistry. Problems. 992pp. 5³/₈ x 8¹/₂. 0-486-65622-5

ELECTRON CORRELATION IN MOLECULES, S. Wilson. This text addresses one of theoretical chemistry's central problems. Topics include molecular electronic structure, independent electron models, electron correlation, the linked diagram theorem, and related topics. 1984 edition. 304pp. 5³/₈ x 8¹/₂. 0-486-45879-2

Engineering

DE RE METALLICA, Georgius Agricola. The famous Hoover translation of greatest treatise on technological chemistry, engineering, geology, mining of early modern times (1556). All 289 original woodcuts. 638pp. $6^3/_4$ x 11. 0-486-60006-8

FUNDAMENTALS OF ASTRODYNAMICS, Roger Bate et al. Modern approach developed by U.S. Air Force Academy. Designed as a first course. Problems, exercises. Numerous illustrations. 455pp. $5^3/_8$ x $8^1/_2$. 0-486-60061-0

DYNAMICS OF FLUIDS IN POROUS MEDIA, Jacob Bear. For advanced students of ground water hydrology, soil mechanics and physics, drainage and irrigation engineering and more. 335 illustrations. Exercises, with answers. 784pp. $6^1/_8$ x $9^1/_4$. 0-486-65675-6

THEORY OF VISCOELASTICITY (SECOND EDITION), Richard M. Christensen. Complete consistent description of the linear theory of the viscoelastic behavior of materials. Problem-solving techniques discussed. 1982 edition. 29 figures. xiv+364pp. $6^1/_8$ x $9^1/_4$. 0-486-42880-X

MECHANICS, J. P. Den Hartog. A classic introductory text or refresher. Hundreds of applications and design problems illuminate fundamentals of trusses, loaded beams and cables, etc. 334 answered problems. 462pp. $5^3/_8$ x $8^1/_2$. 0-486-60754-2

MECHANICAL VIBRATIONS, J. P. Den Hartog. Classic textbook offers lucid explanations and illustrative models, applying theories of vibrations to a variety of practical industrial engineering problems. Numerous figures. 233 problems, solutions. Appendix. Index. Preface. 436pp. $5^3/_8$ x $8^1/_2$. 0-486-64785-4

STRENGTH OF MATERIALS, J. P. Den Hartog. Full, clear treatment of basic material (tension, torsion, bending, etc.) plus advanced material on engineering methods, applications. 350 answered problems. 323pp. $5^3/_8$ x $8^1/_2$. 0-486-60755-0

A HISTORY OF MECHANICS, René Dugas. Monumental study of mechanical principles from antiquity to quantum mechanics. Contributions of ancient Greeks, Galileo, Leonardo, Kepler, Lagrange, many others. 671pp. $5^3/_8$ x $8^1/_2$. 0-486-65632-2

STABILITY THEORY AND ITS APPLICATIONS TO STRUCTURAL MECHANICS, Clive L. Dym. Self-contained text focuses on Koiter postbuckling analyses, with mathematical notions of stability of motion. Basing minimum energy principles for static stability upon dynamic concepts of stability of motion, it develops asymptotic buckling and postbuckling analyses from potential energy considerations, with applications to columns, plates, and arches. 1974 ed. 208pp. $5^3/_8$ x $8^1/_2$. 0-486-42541-X

BASIC ELECTRICITY, U.S. Bureau of Naval Personnel. Originally a training course; best nontechnical coverage. Topics include batteries, circuits, conductors, AC and DC, inductance and capacitance, generators, motors, transformers, amplifiers, etc. Many questions with answers. 349 illustrations. 1969 edition. 448pp. $6^1/_2$ x $9^1/_4$. 0-486-20973-3

ROCKETS, Robert Goddard. Two of the most significant publications in the history of rocketry and jet propulsion: "A Method of Reaching Extreme Altitudes" (1919) and "Liquid Propellant Rocket Development" (1936). 128pp. $5^3/_8 \times 8^1/_2$. 0-486-42537-1

STATISTICAL MECHANICS: PRINCIPLES AND APPLICATIONS, Terrell L. Hill. Standard text covers fundamentals of statistical mechanics, applications to fluctuation theory, imperfect gases, distribution functions, more. 448pp. $5^3/_8 \times 8^1/_2$. 0-486-65390-0

ENGINEERING AND TECHNOLOGY 1650–1750: ILLUSTRATIONS AND TEXTS FROM ORIGINAL SOURCES, Martin Jensen. Highly readable text with more than 200 contemporary drawings and detailed engravings of engineering projects dealing with surveying, leveling, materials, hand tools, lifting equipment, transport and erection, piling, bailing, water supply, hydraulic engineering, and more. Among the specific projects outlined-transporting a 50-ton stone to the Louvre, erecting an obelisk, building timber locks, and dredging canals. 207pp. $8^3/_8 \times 11^1/_4$. 0-486-42232-1

THE VARIATIONAL PRINCIPLES OF MECHANICS, Cornelius Lanczos. Graduate level coverage of calculus of variations, equations of motion, relativistic mechanics, more. First inexpensive paperbound edition of classic treatise. Index. Bibliography. 418pp. $5^3/_8 \times 8^1/_2$. 0-486-65067-7

PROTECTION OF ELECTRONIC CIRCUITS FROM OVERVOLTAGES, Ronald B. Standler. Five-part treatment presents practical rules and strategies for circuits designed to protect electronic systems from damage by transient overvoltages. 1989 ed. xxiv+434pp. $6^1/_8 \times 9^1/_4$. 0-486-42552-5

ROTARY WING AERODYNAMICS, W. Z. Stepniewski. Clear, concise text covers aerodynamic phenomena of the rotor and offers guidelines for helicopter performance evaluation. Originally prepared for NASA. 537 figures. 640pp. $6^1/_8 \times 9^1/_4$. 0-486-64647-5

INTRODUCTION TO SPACE DYNAMICS, William Tyrrell Thomson. Comprehensive, classic introduction to space-flight engineering for advanced undergraduate and graduate students. Includes vector algebra, kinematics, transformation of coordinates. Bibliography. Index. 352pp. $5^3/_8 \times 8^1/_2$. 0-486-65113-4

HISTORY OF STRENGTH OF MATERIALS, Stephen P. Timoshenko. Excellent historical survey of the strength of materials with many references to the theories of elasticity and structure. 245 figures. 452pp. $5^3/_8 \times 8^1/_2$. 0-486-61187-6

ANALYTICAL FRACTURE MECHANICS, David J. Unger. Self-contained text supplements standard fracture mechanics texts by focusing on analytical methods for determining crack-tip stress and strain fields. 336pp. $6^1/_8 \times 9^1/_4$. 0-486-41737-9

STATISTICAL MECHANICS OF ELASTICITY, J. H. Weiner. Advanced, self-contained treatment illustrates general principles and elastic behavior of solids. Part 1, based on classical mechanics, studies thermoelastic behavior of crystalline and polymeric solids. Part 2, based on quantum mechanics, focuses on interatomic force laws, behavior of solids, and thermally activated processes. For students of physics and chemistry and for polymer physicists. 1983 ed. 96 figures. 496pp. $5^3/_8 \times 8^1/_2$. 0-486-42260-7

Mathematics

FUNCTIONAL ANALYSIS (Second Corrected Edition), George Bachman and Lawrence Narici. Excellent treatment of subject geared toward students with background in linear algebra, advanced calculus, physics and engineering. Text covers introduction to inner-product spaces, normed, metric spaces, and topological spaces; complete orthonormal sets, the Hahn-Banach Theorem and its consequences, and many other related subjects. 1966 ed. 544pp. $6^1/_8$ x $9^1/_4$. 0-486-40251-7

DIFFERENTIAL MANIFOLDS, Antoni A. Kosinski. Introductory text for advanced undergraduates and graduate students presents systematic study of the topological structure of smooth manifolds, starting with elements of theory and concluding with method of surgery. 1993 edition. 288pp. $5^3/_8$ x $8^1/_2$. 0-486-46244-7

VECTOR AND TENSOR ANALYSIS WITH APPLICATIONS, A. I. Borisenko and I. E. Tarapov. Concise introduction. Worked-out problems, solutions, exercises. 257pp. $5^5/_8$ x $8^1/_4$. 0-486-63833-2

AN INTRODUCTION TO ORDINARY DIFFERENTIAL EQUATIONS, Earl A. Coddington. A thorough and systematic first course in elementary differential equations for undergraduates in mathematics and science, with many exercises and problems (with answers). Index. 304pp. $5^3/_8$ x $8^1/_2$. 0-486-65942-9

FOURIER SERIES AND ORTHOGONAL FUNCTIONS, Harry F. Davis. An incisive text combining theory and practical example to introduce Fourier series, orthogonal functions and applications of the Fourier method to boundary-value problems. 570 exercises. Answers and notes. 416pp. $5^3/_8$ x $8^1/_2$. 0-486-65973-9

COMPUTABILITY AND UNSOLVABILITY, Martin Davis. Classic graduate-level introduction to theory of computability, usually referred to as theory of recurrent functions. New preface and appendix. 288pp. $5^3/_8$ x $8^1/_2$. 0-486-61471-9

AN INTRODUCTION TO MATHEMATICAL ANALYSIS, Robert A. Rankin. Dealing chiefly with functions of a single real variable, this text by a distinguished educator introduces limits, continuity, differentiability, integration, convergence of infinite series, double series, and infinite products. 1963 edition. 624pp. $5^3/_8$ x $8^1/_2$. 0-486-46251-X

METHODS OF NUMERICAL INTEGRATION (SECOND EDITION), Philip J. Davis and Philip Rabinowitz. Requiring only a background in calculus, this text covers approximate integration over finite and infinite intervals, error analysis, approximate integration in two or more dimensions, and automatic integration. 1984 edition. 624pp. $5^3/_8$ x $8^1/_2$. 0-486-45339-1

INTRODUCTION TO LINEAR ALGEBRA AND DIFFERENTIAL EQUATIONS, John W. Dettman. Excellent text covers complex numbers, determinants, orthonormal bases, Laplace transforms, much more. Exercises with solutions. Undergraduate level. 416pp. $5^3/_8$ x $8^1/_2$. 0-486-65191-6

RIEMANN'S ZETA FUNCTION, H. M. Edwards. Superb, high-level study of landmark 1859 publication entitled "On the Number of Primes Less Than a Given Magnitude" traces developments in mathematical theory that it inspired. xiv+315pp. $5^3/_8$ x $8^1/_2$. 0-486-41740-9

CALCULUS OF VARIATIONS WITH APPLICATIONS, George M. Ewing. Applications-oriented introduction to variational theory develops insight and promotes understanding of specialized books, research papers. Suitable for advanced undergraduate/graduate students as primary, supplementary text. 352pp. $5^3/_8$ x $8^1/_2$. 0-486-64856-7

MATHEMATICIAN'S DELIGHT, W. W. Sawyer. "Recommended with confidence" by *The Times Literary Supplement,* this lively survey was written by a renowned teacher. It starts with arithmetic and algebra, gradually proceeding to trigonometry and calculus. 1943 edition. 240pp. $5^3/_8$ x $8^1/_2$. 0-486-46240-4

ADVANCED EUCLIDEAN GEOMETRY, Roger A. Johnson. This classic text explores the geometry of the triangle and the circle, concentrating on extensions of Euclidean theory, and examining in detail many relatively recent theorems. 1929 edition. 336pp. $5^3/_8$ x $8^1/_2$. 0-486-46237-4

COUNTEREXAMPLES IN ANALYSIS, Bernard R. Gelbaum and John M. H. Olmsted. These counterexamples deal mostly with the part of analysis known as "real variables." The first half covers the real number system, and the second half encompasses higher dimensions. 1962 edition. xxiv+198pp. $5^3/_8$ x $8^1/_2$. 0-486-42875-3

CATASTROPHE THEORY FOR SCIENTISTS AND ENGINEERS, Robert Gilmore. Advanced-level treatment describes mathematics of theory grounded in the work of Poincaré, R. Thom, other mathematicians. Also important applications to problems in mathematics, physics, chemistry and engineering. 1981 edition. References. 28 tables. 397 black-and-white illustrations. xvii + 666pp. $6^1/_8$ x $9^1/_4$. 0-486-67539-4

COMPLEX VARIABLES: Second Edition, Robert B. Ash and W. P. Novinger. Suitable for advanced undergraduates and graduate students, this newly revised treatment covers Cauchy theorem and its applications, analytic functions, and the prime number theorem. Numerous problems and solutions. 2004 edition. 224pp. $6^1/_2$ x $9^1/_4$. 0-486-46250-1

NUMERICAL METHODS FOR SCIENTISTS AND ENGINEERS, Richard Hamming. Classic text stresses frequency approach in coverage of algorithms, polynomial approximation, Fourier approximation, exponential approximation, other topics. Revised and enlarged 2nd edition. 721pp. $5^3/_8$ x $8^1/_2$. 0-486-65241-6

INTRODUCTION TO NUMERICAL ANALYSIS (2nd Edition), F. B. Hildebrand. Classic, fundamental treatment covers computation, approximation, interpolation, numerical differentiation and integration, other topics. 150 new problems. 669pp. $5^3/_8$ x $8^1/_2$. 0-486-65363-3

MARKOV PROCESSES AND POTENTIAL THEORY, Robert M. Blumental and Ronald K. Getoor. This graduate-level text explores the relationship between Markov processes and potential theory in terms of excessive functions, multiplicative functionals and subprocesses, additive functionals and their potentials, and dual processes. 1968 edition. 320pp. $5^3/_8$ x $8^1/_2$. 0-486-46263-3

ABSTRACT SETS AND FINITE ORDINALS: An Introduction to the Study of Set Theory, G. B. Keene. This text unites logical and philosophical aspects of set theory in a manner intelligible to mathematicians without training in formal logic and to logicians without a mathematical background. 1961 edition. 112pp. $5^3/_8$ x $8^1/_2$. 0-486-46249-8

INTRODUCTORY REAL ANALYSIS, A.N. Kolmogorov, S. V. Fomin. Translated by Richard A. Silverman. Self-contained, evenly paced introduction to real and functional analysis. Some 350 problems. 403pp. $5^3/_8$ x $8^1/_2$. 0-486-61226-0

APPLIED ANALYSIS, Cornelius Lanczos. Classic work on analysis and design of finite processes for approximating solution of analytical problems. Algebraic equations, matrices, harmonic analysis, quadrature methods, much more. 559pp. $5^3/_8$ x $8^1/_2$. 0-486-65656-X

AN INTRODUCTION TO ALGEBRAIC STRUCTURES, Joseph Landin. Superb self-contained text covers "abstract algebra": sets and numbers, theory of groups, theory of rings, much more. Numerous well-chosen examples, exercises. 247pp. $5^3/_8$ x $8^1/_2$. 0-486-65940-2

QUALITATIVE THEORY OF DIFFERENTIAL EQUATIONS, V. V. Nemytskii and V.V. Stepanov. Classic graduate-level text by two prominent Soviet mathematicians covers classical differential equations as well as topological dynamics and ergodic theory. Bibliographies. 523pp. $5^3/_8$ x $8^1/_2$. 0-486-65954-2

THEORY OF MATRICES, Sam Perlis. Outstanding text covering rank, nonsingularity and inverses in connection with the development of canonical matrices under the relation of equivalence, and without the intervention of determinants. Includes exercises. 237pp. $5^3/_8$ x $8^1/_2$. 0-486-66810-X

INTRODUCTION TO ANALYSIS, Maxwell Rosenlicht. Unusually clear, accessible coverage of set theory, real number system, metric spaces, continuous functions, Riemann integration, multiple integrals, more. Wide range of problems. Undergraduate level. Bibliography. 254pp. $5^3/_8$ x $8^1/_2$. 0-486-65038-3

MODERN NONLINEAR EQUATIONS, Thomas L. Saaty. Emphasizes practical solution of problems; covers seven types of equations. ". . . a welcome contribution to the existing literature. . . ."—*Math Reviews*. 490pp. $5^3/_8$ x $8^1/_2$. 0-486-64232-1

MATRICES AND LINEAR ALGEBRA, Hans Schneider and George Phillip Barker. Basic textbook covers theory of matrices and its applications to systems of linear equations and related topics such as determinants, eigenvalues and differential equations. Numerous exercises. 432pp. $5^3/_8$ x $8^1/_2$. 0-486-66014-1

LINEAR ALGEBRA, Georgi E. Shilov. Determinants, linear spaces, matrix algebras, similar topics. For advanced undergraduates, graduates. Silverman translation. 387pp. $5^3/_8$ x $8^1/_2$. 0-486-63518-X

MATHEMATICAL METHODS OF GAME AND ECONOMIC THEORY: Revised Edition, Jean-Pierre Aubin. This text begins with optimization theory and convex analysis, followed by topics in game theory and mathematical economics, and concluding with an introduction to nonlinear analysis and control theory. 1982 edition. 656pp. $6^1/_8$ x $9^1/_4$. 0-486-46265-X

SET THEORY AND LOGIC, Robert R. Stoll. Lucid introduction to unified theory of mathematical concepts. Set theory and logic seen as tools for conceptual understanding of real number system. 496pp. $5^3/_8$ x $8^1/_4$. 0-486-63829-4

TENSOR CALCULUS, J.L. Synge and A. Schild. Widely used introductory text covers spaces and tensors, basic operations in Riemannian space, non-Riemannian spaces, etc. 324pp. $5^3/_8 \times 8^1/_4$. 0-486-63612-7

ORDINARY DIFFERENTIAL EQUATIONS, Morris Tenenbaum and Harry Pollard. Exhaustive survey of ordinary differential equations for undergraduates in mathematics, engineering, science. Thorough analysis of theorems. Diagrams. Bibliography. Index. 818pp. $5^3/_8 \times 8^1/_2$. 0-486-64940-7

INTEGRAL EQUATIONS, F. G. Tricomi. Authoritative, well-written treatment of extremely useful mathematical tool with wide applications. Volterra Equations, Fredholm Equations, much more. Advanced undergraduate to graduate level. Exercises. Bibliography. 238pp. $5^3/_8 \times 8^1/_2$. 0-486-64828-1

FOURIER SERIES, Georgi P. Tolstov. Translated by Richard A. Silverman. A valuable addition to the literature on the subject, moving clearly from subject to subject and theorem to theorem. 107 problems, answers. 336pp. $5^3/_8 \times 8^1/_2$. 0-486-63317-9

INTRODUCTION TO MATHEMATICAL THINKING, Friedrich Waismann. Examinations of arithmetic, geometry, and theory of integers; rational and natural numbers; complete induction; limit and point of accumulation; remarkable curves; complex and hypercomplex numbers, more. 1959 ed. 27 figures. xii+260pp. $5^3/_8 \times 8^1/_2$. 0-486-42804-8

THE RADON TRANSFORM AND SOME OF ITS APPLICATIONS, Stanley R. Deans. Of value to mathematicians, physicists, and engineers, this excellent introduction covers both theory and applications, including a rich array of examples and literature. Revised and updated by the author. 1993 edition. 304pp. $6^1/_8 \times 9^1/_4$. 0-486-46241-2

CALCULUS OF VARIATIONS, Robert Weinstock. Basic introduction covering isoperimetric problems, theory of elasticity, quantum mechanics, electrostatics, etc. Exercises throughout. 326pp. $5^3/_8 \times 8^1/_2$. 0-486-63069-2

THE CONTINUUM: A CRITICAL EXAMINATION OF THE FOUNDATION OF ANALYSIS, Hermann Weyl. Classic of 20th-century foundational research deals with the conceptual problem posed by the continuum. 156pp. $5^3/_8 \times 8^1/_2$. 0-486-67982-9

CHALLENGING MATHEMATICAL PROBLEMS WITH ELEMENTARY SOLUTIONS, A. M. Yaglom and I. M. Yaglom. Over 170 challenging problems on probability theory, combinatorial analysis, points and lines, topology, convex polygons, many other topics. Solutions. Total of 445pp. $5^3/_8 \times 8^1/_2$. Two-vol. set.
Vol. I: 0-486-65536-9 Vol. II: 0-486-65537-7

INTRODUCTION TO PARTIAL DIFFERENTIAL EQUATIONS WITH APPLICATIONS, E. C. Zachmanoglou and Dale W. Thoe. Essentials of partial differential equations applied to common problems in engineering and the physical sciences. Problems and answers. 416pp. $5^3/_8 \times 8^1/_2$. 0-486-65251-3

STOCHASTIC PROCESSES AND FILTERING THEORY, Andrew H. Jazwinski. This unified treatment presents material previously available only in journals, and in terms accessible to engineering students. Although theory is emphasized, it discusses numerous practical applications as well. 1970 edition. 400pp. $5^3/_8 \times 8^1/_2$. 0-486-46274-9

Math—Decision Theory, Statistics, Probability

INTRODUCTION TO PROBABILITY, John E. Freund. Featured topics include permutations and factorials, probabilities and odds, frequency interpretation, mathematical expectation, decision-making, postulates of probability, rule of elimination, much more. Exercises with some solutions. Summary. 1973 edition. 247pp. 5³/₈ x 8¹/₂. 0-486-67549-1

STATISTICAL AND INDUCTIVE PROBABILITIES, Hugues Leblanc. This treatment addresses a decades-old dispute among probability theorists, asserting that both statistical and inductive probabilities may be treated as sentence-theoretic measurements, and that the latter qualify as estimates of the former. 1962 edition. 160pp. 5³/₈ x 8¹/₂. 0-486-44980-7

APPLIED MULTIVARIATE ANALYSIS: Using Bayesian and Frequentist Methods of Inference, Second Edition, S. James Press. This two-part treatment deals with foundations as well as models and applications. Topics include continuous multivariate distributions; regression and analysis of variance; factor analysis and latent structure analysis; and structuring multivariate populations. 1982 edition. 692pp. 5³/₈ x 8¹/₂. 0-486-44236-5

LINEAR PROGRAMMING AND ECONOMIC ANALYSIS, Robert Dorfman, Paul A. Samuelson and Robert M. Solow. First comprehensive treatment of linear programming in standard economic analysis. Game theory, modern welfare economics, Leontief input-output, more. 525pp. 5³/₈ x 8¹/₂. 0-486-65491-5

PROBABILITY: AN INTRODUCTION, Samuel Goldberg. Excellent basic text covers set theory, probability theory for finite sample spaces, binomial theorem, much more. 360 problems. Bibliographies. 322pp. 5³/₈ x 8¹/₂. 0-486-65252-1

GAMES AND DECISIONS: INTRODUCTION AND CRITICAL SURVEY, R. Duncan Luce and Howard Raiffa. Superb nontechnical introduction to game theory, primarily applied to social sciences. Utility theory, zero-sum games, n-person games, decision-making, much more. Bibliography. 509pp. 5³/₈ x 8¹/₂. 0-486-65943-7

INTRODUCTION TO THE THEORY OF GAMES, J. C. C. McKinsey. This comprehensive overview of the mathematical theory of games illustrates applications to situations involving conflicts of interest, including economic, social, political, and military contexts. Appropriate for advanced undergraduate and graduate courses; advanced calculus a prerequisite. 1952 ed. x+372pp. 5³/₈ x 8¹/₂. 0-486-42811-7

FIFTY CHALLENGING PROBLEMS IN PROBABILITY WITH SOLUTIONS, Frederick Mosteller. Remarkable puzzlers, graded in difficulty, illustrate elementary and advanced aspects of probability. Detailed solutions. 88pp. 5³/₈ x 8¹/₂. 0-486-65355-2

PROBABILITY THEORY: A CONCISE COURSE, Y. A. Rozanov. Highly readable, self-contained introduction covers combination of events, dependent events, Bernoulli trials, etc. 148pp. 5³/₈ x 8¹/₄. 0-486-63544-9

THE STATISTICAL ANALYSIS OF EXPERIMENTAL DATA, John Mandel. First half of book presents fundamental mathematical definitions, concepts and facts while remaining half deals with statistics primarily as an interpretive tool. Well-written text, numerous worked examples with step-by-step presentation. Includes 116 tables. 448pp. 5³/₈ x 8¹/₂. 0-486-64666-1

Math—Geometry and Topology

ELEMENTARY CONCEPTS OF TOPOLOGY, Paul Alexandroff. Elegant, intuitive approach to topology from set-theoretic topology to Betti groups; how concepts of topology are useful in math and physics. 25 figures. 57pp. $5^3/_8$ x $8^1/_2$. 0-486-60747-X

A LONG WAY FROM EUCLID, Constance Reid. Lively guide by a prominent historian focuses on the role of Euclid's Elements in subsequent mathematical developments. Elementary algebra and plane geometry are sole prerequisites. 80 drawings. 1963 edition. 304pp. $5^3/_8$ x $8^1/_2$. 0-486-43613-6

EXPERIMENTS IN TOPOLOGY, Stephen Barr. Classic, lively explanation of one of the byways of mathematics. Klein bottles, Moebius strips, projective planes, map coloring, problem of the Koenigsberg bridges, much more, described with clarity and wit. 43 figures. 210pp. $5^3/_8$ x $8^1/_2$. 0-486-25933-1

THE GEOMETRY OF RENÉ DESCARTES, René Descartes. The great work founded analytical geometry. Original French text, Descartes's own diagrams, together with definitive Smith-Latham translation. 244pp. $5^3/_8$ x $8^1/_2$. 0-486-60068-8

EUCLIDEAN GEOMETRY AND TRANSFORMATIONS, Clayton W. Dodge. This introduction to Euclidean geometry emphasizes transformations, particularly isometries and similarities. Suitable for undergraduate courses, it includes numerous examples, many with detailed answers. 1972 ed. viii+296pp. $6^1/_8$ x $9^1/_4$. 0-486-43476-1

EXCURSIONS IN GEOMETRY, C. Stanley Ogilvy. A straightedge, compass, and a little thought are all that's needed to discover the intellectual excitement of geometry. Harmonic division and Apollonian circles, inversive geometry, hexlet, Golden Section, more. 132 illustrations. 192pp. $5^3/_8$ x $8^1/_2$. 0-486-26530-7

THE THIRTEEN BOOKS OF EUCLID'S ELEMENTS, translated with introduction and commentary by Sir Thomas L. Heath. Definitive edition. Textual and linguistic notes, mathematical analysis. 2,500 years of critical commentary. Unabridged. 1,414pp. $5^3/_8$ x $8^1/_2$. Three-vol. set.
Vol. I: 0-486-60088-2 Vol. II: 0-486-60089-0 Vol. III: 0-486-60090-4

SPACE AND GEOMETRY: IN THE LIGHT OF PHYSIOLOGICAL, PSYCHOLOGICAL AND PHYSICAL INQUIRY, Ernst Mach. Three essays by an eminent philosopher and scientist explore the nature, origin, and development of our concepts of space, with a distinctness and precision suitable for undergraduate students and other readers. 1906 ed. vi+148pp. $5^3/_8$ x $8^1/_2$. 0-486-43909-7

GEOMETRY OF COMPLEX NUMBERS, Hans Schwerdtfeger. Illuminating, widely praised book on analytic geometry of circles, the Moebius transformation, and two-dimensional non-Euclidean geometries. 200pp. $5^3/_8$ x $8^1/_4$. 0-486-63830-8

DIFFERENTIAL GEOMETRY, Heinrich W. Guggenheimer. Local differential geometry as an application of advanced calculus and linear algebra. Curvature, transformation groups, surfaces, more. Exercises. 62 figures. 378pp. $5^3/_8$ x $8^1/_2$. 0-486-63433-7

History of Math

THE WORKS OF ARCHIMEDES, Archimedes (T. L. Heath, ed.). Topics include the famous problems of the ratio of the areas of a cylinder and an inscribed sphere; the measurement of a circle; the properties of conoids, spheroids, and spirals; and the quadrature of the parabola. Informative introduction. clxxxvi+326pp. $5^3/_8$ x $8^1/_2$. 0-486-42084-1

A SHORT ACCOUNT OF THE HISTORY OF MATHEMATICS, W. W. Rouse Ball. One of clearest, most authoritative surveys from the Egyptians and Phoenicians through 19th-century figures such as Grassman, Galois, Riemann. Fourth edition. 522pp. $5^3/_8$ x $8^1/_2$. 0-486-20630-0

THE HISTORY OF THE CALCULUS AND ITS CONCEPTUAL DEVELOPMENT, Carl B. Boyer. Origins in antiquity, medieval contributions, work of Newton, Leibniz, rigorous formulation. Treatment is verbal. 346pp. $5^3/_8$ x $8^1/_2$. 0-486-60509-4

THE HISTORICAL ROOTS OF ELEMENTARY MATHEMATICS, Lucas N. H. Bunt, Phillip S. Jones, and Jack D. Bedient. Fundamental underpinnings of modern arithmetic, algebra, geometry and number systems derived from ancient civilizations. 320pp. $5^3/_8$ x $8^1/_2$. 0-486-25563-8

THE HISTORY OF THE CALCULUS AND ITS CONCEPTUAL DEVELOPMENT, Carl B. Boyer. Fluent description of the development of both the integral and differential calculus—its early beginnings in antiquity, medieval contributions, and a consideration of Newton and Leibniz. 368pp. $5^3/_8$ x $8^1/_2$. 0-486-60509-4

GAMES, GODS & GAMBLING: A HISTORY OF PROBABILITY AND STATISTICAL IDEAS, F. N. David. Episodes from the lives of Galileo, Fermat, Pascal, and others illustrate this fascinating account of the roots of mathematics. Features thought-provoking references to classics, archaeology, biography, poetry. 1962 edition. 304pp. $5^3/_8$ x $8^1/_2$. (Available in U.S. only.) 0-486-40023-9

OF MEN AND NUMBERS: THE STORY OF THE GREAT MATHEMATICIANS, Jane Muir. Fascinating accounts of the lives and accomplishments of history's greatest mathematical minds—Pythagoras, Descartes, Euler, Pascal, Cantor, many more. Anecdotal, illuminating. 30 diagrams. Bibliography. 256pp. $5^3/_8$ x $8^1/_2$. 0-486-28973-7

HISTORY OF MATHEMATICS, David E. Smith. Nontechnical survey from ancient Greece and Orient to late 19th century; evolution of arithmetic, geometry, trigonometry, calculating devices, algebra, the calculus. 362 illustrations. 1,355pp. $5^3/_8$ x $8^1/_2$. Two-vol. set. Vol. I: 0-486-20429-4 Vol. II: 0-486-20430-8

A CONCISE HISTORY OF MATHEMATICS, Dirk J. Struik. The best brief history of mathematics. Stresses origins and covers every major figure from ancient Near East to 19th century. 41 illustrations. 195pp. $5^3/_8$ x $8^1/_2$. 0-486-60255-9

Physics

OPTICAL RESONANCE AND TWO-LEVEL ATOMS, L. Allen and J. H. Eberly. Clear, comprehensive introduction to basic principles behind all quantum optical resonance phenomena. 53 illustrations. Preface. Index. 256pp. 5³/₈ x 8¹/₂. 0-486-65533-4

QUANTUM THEORY, David Bohm. This advanced undergraduate-level text presents the quantum theory in terms of qualitative and imaginative concepts, followed by specific applications worked out in mathematical detail. Preface. Index. 655pp. 5³/₈ x 8¹/₂. 0-486-65969-0

ATOMIC PHYSICS (8th EDITION), Max Born. Nobel laureate's lucid treatment of kinetic theory of gases, elementary particles, nuclear atom, wave-corpuscles, atomic structure and spectral lines, much more. Over 40 appendices, bibliography. 495pp. 5³/₈ x 8¹/₂. 0-486-65984-4

A SOPHISTICATE'S PRIMER OF RELATIVITY, P. W. Bridgman. Geared toward readers already acquainted with special relativity, this book transcends the view of theory as a working tool to answer natural questions: What is a frame of reference? What is a "law of nature"? What is the role of the "observer"? Extensive treatment, written in terms accessible to those without a scientific background. 1983 ed. xlviii+172pp. 5³/₈ x 8¹/₂. 0-486-42549-5

AN INTRODUCTION TO HAMILTONIAN OPTICS, H. A. Buchdahl. Detailed account of the Hamiltonian treatment of aberration theory in geometrical optics. Many classes of optical systems defined in terms of the symmetries they possess. Problems with detailed solutions. 1970 edition. xv + 360pp. 5³/₈ x 8¹/₂. 0-486-67597-1

PRIMER OF QUANTUM MECHANICS, Marvin Chester. Introductory text examines the classical quantum bead on a track: its state and representations; operator eigenvalues; harmonic oscillator and bound bead in a symmetric force field; and bead in a spherical shell. Other topics include spin, matrices, and the structure of quantum mechanics; the simplest atom; indistinguishable particles; and stationary-state perturbation theory. 1992 ed. xiv+314pp. 6¹/₈ x 9¹/₄. 0-486-42878-8

LECTURES ON QUANTUM MECHANICS, Paul A. M. Dirac. Four concise, brilliant lectures on mathematical methods in quantum mechanics from Nobel Prize-winning quantum pioneer build on idea of visualizing quantum theory through the use of classical mechanics. 96pp. 5³/₈ x 8¹/₂. 0-486-41713-1

THIRTY YEARS THAT SHOOK PHYSICS: THE STORY OF QUANTUM THEORY, George Gamow. Lucid, accessible introduction to influential theory of energy and matter. Careful explanations of Dirac's anti-particles, Bohr's model of the atom, much more. 12 plates. Numerous drawings. 240pp. 5³/₈ x 8¹/₂. 0-486-24895-X

ELECTRONIC STRUCTURE AND THE PROPERTIES OF SOLIDS: THE PHYSICS OF THE CHEMICAL BOND, Walter A. Harrison. Innovative text offers basic understanding of the electronic structure of covalent and ionic solids, simple metals, transition metals and their compounds. Problems. 1980 edition. 582pp. 6¹/₈ x 9¹/₄. 0-486-66021-4

HYDRODYNAMIC AND HYDROMAGNETIC STABILITY, S. Chandrasekhar. Lucid examination of the Rayleigh-Benard problem; clear coverage of the theory of instabilities causing convection. 704pp. $5^3/_8$ x $8^1/_4$. 0-486-64071-X

INVESTIGATIONS ON THE THEORY OF THE BROWNIAN MOVEMENT, Albert Einstein. Five papers (1905–8) investigating dynamics of Brownian motion and evolving elementary theory. Notes by R. Fürth. 122pp. $5^3/_8$ x $8^1/_2$. 0-486-60304-0

THE PHYSICS OF WAVES, William C. Elmore and Mark A. Heald. Unique overview of classical wave theory. Acoustics, optics, electromagnetic radiation, more. Ideal as classroom text or for self-study. Problems. 477pp. $5^3/_8$ x $8^1/_2$. 0-486-64926-1

GRAVITY, George Gamow. Distinguished physicist and teacher takes reader-friendly look at three scientists whose work unlocked many of the mysteries behind the laws of physics: Galileo, Newton, and Einstein. Most of the book focuses on Newton's ideas, with a concluding chapter on post-Einsteinian speculations concerning the relationship between gravity and other physical phenomena. 160pp. $5^3/_8$ x $8^1/_2$. 0-486-42563-0

PHYSICAL PRINCIPLES OF THE QUANTUM THEORY, Werner Heisenberg. Nobel Laureate discusses quantum theory, uncertainty, wave mechanics, work of Dirac, Schroedinger, Compton, Wilson, Einstein, etc. 184pp. $5^3/_8$ x $8^1/_2$. 0-486-60113-7

ATOMIC SPECTRA AND ATOMIC STRUCTURE, Gerhard Herzberg. One of best introductions; especially for specialist in other fields. Treatment is physical rather than mathematical. 80 illustrations. 257pp. $5^3/_8$ x $8^1/_2$. 0-486-60115-3

AN INTRODUCTION TO STATISTICAL THERMODYNAMICS, Terrell L. Hill. Excellent basic text offers wide-ranging coverage of quantum statistical mechanics, systems of interacting molecules, quantum statistics, more. 523pp. $5^3/_8$ x $8^1/_2$. 0-486-65242-4

THEORETICAL PHYSICS, Georg Joos, with Ira M. Freeman. Classic overview covers essential math, mechanics, electromagnetic theory, thermodynamics, quantum mechanics, nuclear physics, other topics. First paperback edition. xxiii + 885pp. $5^3/_8$ x $8^1/_2$. 0-486-65227-0

PROBLEMS AND SOLUTIONS IN QUANTUM CHEMISTRY AND PHYSICS, Charles S. Johnson, Jr. and Lee G. Pedersen. Unusually varied problems, detailed solutions in coverage of quantum mechanics, wave mechanics, angular momentum, molecular spectroscopy, more. 280 problems plus 139 supplementary exercises. 430pp. $6^1/_2$ x $9^1/_4$. 0-486-65236-X

THEORETICAL SOLID STATE PHYSICS, Vol. 1: Perfect Lattices in Equilibrium; Vol. II: Non-Equilibrium and Disorder, William Jones and Norman H. March. Monumental reference work covers fundamental theory of equilibrium properties of perfect crystalline solids, non-equilibrium properties, defects and disordered systems. Appendices. Problems. Preface. Diagrams. Index. Bibliography. Total of 1,301pp. $5^3/_8$ x $8^1/_2$. Two volumes. Vol. I: 0-486-65015-4 Vol. II: 0-486-65016-2

WHAT IS RELATIVITY? L. D. Landau and G. B. Rumer. Written by a Nobel Prize physicist and his distinguished colleague, this compelling book explains the special theory of relativity to readers with no scientific background, using such familiar objects as trains, rulers, and clocks. 1960 ed. vi+72pp. $5^3/_8$ x $8^1/_2$. 0-486-42806-0

A TREATISE ON ELECTRICITY AND MAGNETISM, James Clerk Maxwell. Important foundation work of modern physics. Brings to final form Maxwell's theory of electromagnetism and rigorously derives his general equations of field theory. 1,084pp. $5^3/_8$ x $8^1/_2$. Two-vol. set. Vol. I: 0-486-60636-8 Vol. II: 0-486-60637-6

MATHEMATICS FOR PHYSICISTS, Philippe Dennery and Andre Krzywicki. Superb text provides math needed to understand today's more advanced topics in physics and engineering. Theory of functions of a complex variable, linear vector spaces, much more. Problems. 1967 edition. 400pp. $6^1/_2$ x $9^1/_4$. 0-486-69193-4

INTRODUCTION TO QUANTUM MECHANICS WITH APPLICATIONS TO CHEMISTRY, Linus Pauling & E. Bright Wilson, Jr. Classic undergraduate text by Nobel Prize winner applies quantum mechanics to chemical and physical problems. Numerous tables and figures enhance the text. Chapter bibliographies. Appendices. Index. 468pp. $5^3/_8$ x $8^1/_2$. 0-486-64871-0

METHODS OF THERMODYNAMICS, Howard Reiss. Outstanding text focuses on physical technique of thermodynamics, typical problem areas of understanding, and significance and use of thermodynamic potential. 1965 edition. 238pp. $5^3/_8$ x $8^1/_2$. 0-486-69445-3

THE ELECTROMAGNETIC FIELD, Albert Shadowitz. Comprehensive under- graduate text covers basics of electric and magnetic fields, builds up to electromagnetic theory. Also related topics, including relativity. Over 900 problems. 768pp. $5^5/_8$ x $8^1/_4$. 0-486-65660-8

GREAT EXPERIMENTS IN PHYSICS: FIRSTHAND ACCOUNTS FROM GALILEO TO EINSTEIN, Morris H. Shamos (ed.). 25 crucial discoveries: Newton's laws of motion, Chadwick's study of the neutron, Hertz on electromagnetic waves, more. Original accounts clearly annotated. 370pp. $5^3/_8$ x $8^1/_2$. 0-486-25346-5

EINSTEIN'S LEGACY, Julian Schwinger. A Nobel Laureate relates fascinating story of Einstein and development of relativity theory in well-illustrated, nontechnical volume. Subjects include meaning of time, paradoxes of space travel, gravity and its effect on light, non-Euclidean geometry and curving of space-time, impact of radio astronomy and space-age discoveries, and more. 189 b/w illustrations. xiv+250pp. $8^3/_8$ x $9^1/_4$. 0-486-41974-6

THE VARIATIONAL PRINCIPLES OF MECHANICS, Cornelius Lanczos. Philosophic, less formalistic approach to analytical mechanics offers model of clear, scholarly exposition at graduate level with coverage of basics, calculus of variations, principle of virtual work, equations of motion, more. 418pp. $5^3/_8$ x $8^1/_2$. 0-486-65067-7

Paperbound unless otherwise indicated. Available at your book dealer, online at www.doverpublications.com, or by writing to Dept. GI, Dover Publications, Inc., 31 East 2nd Street, Mineola, NY 11501. For current price information or for free catalogues (please indicate field of interest), write to Dover Publications or log on to www.doverpublications.com and see every Dover book in print. Dover publishes more than 400 books each year on science, elementary and advanced mathematics, biology, music, art, literary history, social sciences, and other areas.